Springer Optimization and Its Applications

VOLUME 85

Aims and Scope

Optimization has been expanding in all directions at an astonishing rate during the last few decades. New algorithmic and theoretical techniques have been developed, the diffusion into other disciplines has proceeded at a rapid pace, and our knowledge of all aspects of the field has grown even more profound. At the same time, one of the most striking trends in optimization is the constantly increasing emphasis on the interdisciplinary nature of the field. Optimization has been a basic tool in all areas of applied mathematics, engineering, medicine, economics, and other sciences.

The series *Springer Optimization and Its Applications* publishes undergraduate and graduate textbooks, monographs and state-of-the-art expository work that focus on algorithms for solving optimization problems and also study applications involving such problems. Some of the topics covered include nonlinear optimization (convex and nonconvex), network flow problems, stochastic optimization, optimal control, discrete optimization, multi-objective programming, description of software packages, approximation techniques and heuristic approaches.

For further volumes:
http://www.springer.com/series/7393

Michael Zabarankin • Stan Uryasev

Statistical Decision Problems

Selected Concepts and Portfolio Safeguard Case Studies

Springer

Michael Zabarankin
Department of Mathematical Sciences
Stevens Institute of Technology
Hoboken, NJ, USA

Stan Uryasev
Department of Industrial and Systems Engineering
University of Florida
Gainesville, FL, USA

ISSN 1931-6828
ISBN 978-1-4939-5325-7 ISBN 978-1-4614-8471-4 (eBook)
DOI 10.1007/978-1-4614-8471-4
Springer New York Heidelberg Dordrecht London

Mathematics Subject Classification (2010): 62C05, 62J05, 62J12, 62P05, 62P30, 90B50, 90B90, 90C05, 90C11, 90C20, 90C26, 90C90, 91B06, 91B30, 91B70

Printed on acid-free paper

Springer is part of Springer Science+Business Media (www.springer.com)

Preface

Scope

In an abstract form, statistical decision making is an optimization problem that uses available statistical data as an input and optimizes an objective function of interest with respect to decision variables subject to certain constraints. Typically, uncertainty encoded in statistical data can be translated into five basic notions: *likelihood*, *entropy*, *error*, *deviation*, and *risk*. As a result, the majority of statistical decision problems can be tentatively divided into four major categories: (i) likelihood maximization, (ii) entropy maximization (relative entropy minimization), (iii) error minimization (regression), and (iv) decision models in which deviation or risk is either minimized or constrained. All these problems may include so-called technical constrains on decision variables, e.g., box and cardinality constraints. It is also common to optimize one of the corresponding five functionals while to constrain another, e.g., maximizing entropy subject to a constraint on deviation, or to find a trade-off between one of the functionals and the expected value of the quantity of interest. The book aims to demonstrate how to use these "building blocks": likelihood, entropy, error, deviation, and risk to formulate statistical decision problems arising in various risk management applications, e.g., optimal hedging, portfolio optimization, portfolio replication, cash flow matching, and classification and how to solve those problems in optimization package *Portfolio Safeguard* (PSG).

Content

The book consists of three parts: selected concepts of statistical decision theory (Part I), statistical decision problems (Part II), and case studies with PSG (Part III). Part I presents a general theory of error, deviation, and risk measures to be used in various statistical decision problems and also discusses probabilistic inequalities with deviation measures such as generalized Chebyshev's and Kolmogorov's inequalities. Part II covers five major topics: parametric and nonparametric estimation based on the maximum likelihood principle, entropy maximization problems, unconstrained and constrained linear regression with general error measures, classification with logistic regression and with support vector machines (SVMs), and statistical decision problems with general deviation and risk measures. Part III discusses 21 case studies of typical statistical decision problems arising in risk management, particularly in financial engineering, and demonstrates implementation of those problems in PSG. All case studies are closely related to theoretical Part II and are examples of statistical decision problems from the four categories (i)–(iv).

Audience

The book is aimed at practitioners working in the areas of risk management, decision making under uncertainty, and statistics. It can serve as a quick introduction into the theory of general error, deviation, and risk measures for the graduate students, engineers, and statisticians interested in modeling and managing risk in various applications such as optimal hedging, portfolio replication, portfolio optimization, cash flow matching, structuring of collateralized debt obligations (CDOs), classification, sparse signal reconstruction, and therapy treatment optimization, to mention just a few. It can also be used as a supplementary reading for a number of graduate courses including but not limited to those of statistical analysis, models of risk, data mining, stochastic programming, financial engineering, modern portfolio theory, and advanced engineering economy.

Optimization Software: Portfolio Safeguard

PSG is an advanced nonlinear mixed-integer optimization package for solving a wide range of optimization, statistics, and risk management problems. PSG is a product of American Optimal Decisions, Inc. (see www.aorda.com). Although PSG is a general-purpose decision support tool, the focus application areas are risk management, financial engineering, military, and medical applications. PSG is based on a simple but powerful idea: for every engineering area, identify most commonly used nonlinear functions and include them in the package as independent built-in objects. Each function is defined by a function type, parameters, and a matrix of data (e.g., scenario matrix or covariance matrix). Specialized algorithms, built for different types of functions, efficiently optimize large-scale nonlinear functions, such as probability, value-at-risk (VaR), and omega functions, which are typically beyond the scope of commercial packages. The built-in function library provides simple and convenient interface for evaluating functions and their derivatives, for constructing optimization problems and solving them, and for analyzing solutions. No programming experience is required to use PSG.

PSG operates in four programming environments: Shell (Windows), MATLAB, C++, and Run-File (Text). The standard PSG setup includes case studies from various areas with emphasis on financial engineering applications, such as portfolio optimization, asset allocation, selection of insurance, hedging with derivative contracts, bond matching, and structuring of CDOs.

PSG can be downloaded from the American Optimal Decisions web site: www.aorda.com/aod. Four types of licenses are available: Freeware Express, Regular, Academic, and Regular Business. Freeware Express Edition limits the number of decision variables per function to ten. The Regular PSG edition has a free 30-day trial. After installing PSG, case study projects can be viewed in the Case Studies folder in File tab of the PSG menu. In order to modify an existing case study, it should be copied into Work directory. Tutorials about PSG and case study descriptions can be found in Help tab of the menu.

Acknowledgments

We thank Dr. Bogdan Grechuk at the University of Leicester, UK, and Dr. Anton Molyboha at Teza Technologies for carefully proofreading the portions of the manuscript and for providing valuable comments. We are also grateful to Dr. Viktor Kuzmenko at American Optimal Decisions, Inc. for the technical support and assistance with running numerical experiments for the case studies. We express our deepest gratitude to our families for their constant encouragement, support, and patience.

Hoboken, NJ, USA Michael Zabarankin
Gainesville, FL, USA Stan Uryasev

Contents

Nomenclature

$\sim$	"has distribution"
$[x]_+$	$\max\{x, 0\}$ for real-valued x
$\partial\mathscr{F}(X)$	subdifferential of convex functional $\mathscr{F}$ at X
$\succeq_{(n)}$	non-strict preference relation with respect to n^{th}-order stochastic dominance
$\|X\|_p$	p-norm of X in $\mathscr{L}^p$, $p \in [1, \infty]$
$\mathscr{A}$	acceptance set
$\text{AvgDD}(\xi)$	average drawdown of sample path ξ
β_i	beta of asset $i \in \{1, \ldots, n\}$ in the capital asset pricing model (CAPM)
$\mathscr{C}$	set of parameters $c_0, c_1, \ldots, c_n$
$\text{CDaR}_\alpha(\xi)$	conditional drawdown-at-risk (CDaR) of sample path ξ
$\text{Cov}(X, Y)$	covariance of X and Y
$\text{CVaR}_\alpha(X)$	conditional value-at-risk (CVaR) of X
$\text{CVaR}_\alpha^\Delta(X)$	conditional value-at-risk (CVaR) deviation of X
$\text{CVaR}_\alpha^-(L)$	conditional value-at-risk (CVaR) for loss L
$\mathscr{D}(X)$	deviation measure $\mathscr{D}$ of X
$D_{\text{KL}}(X\|Y)$	Kullback–Leibler divergence measure (relative entropy) of X and Y
$\mathscr{E}(X)$	error measure $\mathscr{E}$ of X
$E[X]$	expected value of X
e	n-dimensional unit vector $(1, \ldots, 1)$
$F_X(t)$	cumulative distribution function (CDF) of X
$f_X(t)$	probability density function (PDF) of X
$F_X^{-1}(s)$	inverse of the cumulative distribution function (CDF) of X
$\Phi(t)$	cumulative distribution function (CDF) of the standard normal random variable
$\Gamma(x)$	Gamma function of real-valued x
$H_\alpha(X)$	Renyi entropy of X
$I_{\{\cdot\}}$	indicator function (1 if the condition in the curly brackets is true and 0 otherwise)
inf	essential infimum of X ($\text{ess inf}\, X$)

$\mathscr{L}^p$	Lebesgue space of integrable functions, $p \in [1, \infty]$
Λ	covariance matrix
Λ^{-1}	inverse of covariance matrix Λ
$\mathscr{M}$	sigma-algebra over space of probability events
$M_X(t)$	moment generating function (MGF) of X
$\mathrm{MAD}(X)$	mean absolute deviation (MAD) of X
$\mathrm{MaxDD}(\xi)$	maximum drawdown of sample path ξ
$\mathrm{med}\,(X)$	median of X
μ	mean value of X
μ_M	expected rate of return of master fund (market portfolio)
$\mathbb{N}$	set of positive integer numbers
$N\left(\mu, \sigma^2\right)$	normal distribution with mean μ and variance σ^2
$N_{\mathscr{X}}(X)$	normal cone of nonempty, closed and convex set $\mathscr{X}$ at $X \in \mathscr{X}$
Ω	probability space of elementary events
$\mathbb{P}$	probability measure
$\mathscr{Q}$	risk envelope
$\mathscr{Q}_{\mathscr{D}}(X)$	set of risk identifiers for X with respect to deviation $\mathscr{D}$
$q_X^+(s)$	upper s-quantile of X
$q_X^-(s)$	lower s-quantile of X
$q_X(s)$	s-quantile interval $\left[q_X^-(s), q_X^+(s)\right]$
$\overline{q}_X(s)$	average s-quantile of X
$\mathscr{R}(X)$	risk measure $\mathscr{R}$ of X
$\mathbb{R}$	set of real numbers
$\mathbb{R}^+$	set of positive real numbers
$\mathbb{R}_0^+$	set of nonnegative real numbers
$\mathbb{R}^n$	set of n-dimensional real vectors
r_0	risk-free rate of return
r_i	rate of return of risky asset $i \in \{1, \ldots, n\}$
r_M	rate of return of master fund (market portfolio)
$\mathrm{rc}\mathscr{X}$	recession cone of nonempty, closed and convex set $\mathscr{X}$
$\mathscr{S}(X)$	statistic of X associated with error $\mathscr{E}$
$S(X)$	Shannon entropy of X
$\sigma(X)$	standard deviation of X
$\sigma^2(X)$	variance of X
$\sigma_+(X)$	standard upper semideviation of X
$\sigma_-(X)$	standard lower semideviation of X
σ_M^2	variance of the rate of return of master fund (market portfolio)
sup	essential supremum of X (ess sup X)
$\widehat{\theta}$	statistical estimator of parameter θ
$U(0, 1)$	uniform probability distribution on [0, 1]
$\mathrm{VaR}_\alpha(X)$	value-at-risk (VaR) of X
$\mathrm{VaR}_\alpha^\Delta(X)$	value-at-risk (VaR) deviation of X
X, Y, Z	random variables
$\mathscr{X}$	feasible set of X

Part I
Selected Concepts of Statistical Decision Theory

Chapter 1
Random Variables

Let $(\Omega, \mathcal{M}, \mathbb{P})$ be a probability space of elementary events Ω with the sigma-algebra $\mathcal{M}$ over Ω and with a probability measure $\mathbb{P}$ on $(\Omega, \mathcal{M})$. Random variables are assumed to be measurable real-valued functions from $\mathcal{L}^2(\Omega) = \mathcal{L}^2(\Omega, \mathcal{M}, \mathbb{P})$ unless otherwise specified,[1] and the relationships between random variables X and Y, e.g., $X \leq Y$ and $X = Y$, are understood to hold in the almost sure sense, i.e., $\mathbb{P}[X \leq Y] = 1$ and $\mathbb{P}[X = Y] = 1$, respectively. Also, $\inf X$ and $\sup X$ mean *essential infimum* and *essential supremum* of X, i.e., $\operatorname{ess\,inf} X$ and $\operatorname{ess\,sup} X$, respectively. Two important integral characteristics of a random variable X are its *mean* and *variance* defined by

$$\mu(X) = \int_\Omega X(\omega)\, d\mathbb{P}[\omega], \qquad \sigma^2(X) = \int_\Omega (X(\omega) - \mu(X))^2\, d\mathbb{P}[\omega],$$

respectively, and $\sigma(X)$ is called *standard deviation* of X. If $X \in \mathcal{L}^2(\Omega)$, then $\mu(X)$ and $\sigma^2(X)$ are well defined (bounded), which explains the choice of $\mathcal{L}^2(\Omega)$.

1.1 Probability Distribution Function

The cumulative distribution function (CDF), $F_X(t)$, of a random variable $X : \Omega \mapsto \mathbb{R}$ is defined by $F_X(t) = \mathbb{P}[X \leq t]$. It is nonnegative, nondecreasing, and right continuous and has the following properties: $\lim_{t \to -\infty} F_X(t) = 0$ and $\lim_{t \to \infty} F_X(t) = 1$. If $F_X(t)$ is a differentiable function with respect to t, then $f_X(t) = F_X'(t)$ is the probability density function (PDF) of X. In this case, X is said to be continuously distributed.

---

[1] $\mathcal{L}^2(\Omega)$ is the Lebesgue space of measurable square-integrable functions on Ω: $X \in \mathcal{L}^2(\Omega)$ is equivalent to $\int_\Omega |X(\omega)|^2 d\mathbb{P}[\omega] < \infty$.

M. Zabarankin and S. Uryasev, *Statistical Decision Problems: Selected Concepts and Portfolio Safeguard Case Studies*, Springer Optimization and Its Applications 85,
DOI 10.1007/978-1-4614-8471-4_1,

Similarly, if $X = (X_1, \ldots, X_n)$ is an n-dimensional random vector, then its CDF is defined by $F_X(t_1, \ldots, t_n) = \mathbb{P}[X_1 \le t_1, \ldots, X_n \le t_n]$. In fact, $X_1, \ldots, X_n$ can be considered as a set of random variables, and in this case, the CDF $F_X(t_1, \ldots, t_n)$ is called the *joint* CDF of $X_1, \ldots, X_n$. If $F_X(t_1, \ldots, t_n)$ is continuously differentiable with respect to $t_1, \ldots, t_n$, then $f_X(t_1, \ldots, t_n) = \partial^n F_X(t_1, \ldots, t_n)/(\partial t_1 \ldots \partial t_n)$ is the probability density function of $X_1, \ldots, X_n$, and $X_1, \ldots, X_n$ are said to be continuously distributed. In this case, the PDF of X_k, $k \in \{1, \ldots, n\}$, is determined by the $(n-1)$-dimensional integral

$$f_{X_k}(t_k) = \int_{-\infty}^{\infty} \cdots \int_{-\infty}^{\infty} f_X(t_1, \ldots, t_n)\, dt_1 \ldots dt_{k-1} dt_{k+1} \ldots dt_n$$

and is called the *marginal* PDF of X_k.

Random variables $X_1, \ldots, X_n$ are *independent* or *independently distributed* if their joint CDF is the product of individual CDFs, i.e., if

$$F_X(t_1, \ldots, t_n) = \prod_{k=1}^{n} F_{X_k}(t_k).$$

In particular, continuously distributed $X_1, \ldots, X_n$ with a joint PDF $f_X(t_1, \ldots, t_n)$ are independent if and only if $f_X(t_1, \ldots, t_n) = \prod_{k=1}^{n} f_{X_k}(t_k)$.

A function $g : \mathbb{R} \to \mathbb{R}$ of a random variable X is a random variable $Y = g(X)$ with the CDF $F_Y(t) = \mathbb{P}[Y \le t] = \mathbb{P}[g(X) \le t]$. If g is an increasing function and admits the inverse g^{-1}, then $F_Y(t) = F_X(g^{-1}(t))$. For example, if g is the CDF of X, i.e., $g = F_X$, then Y is uniformly distributed on the interval $[0, 1]$: $\mathbb{P}[Y \le t] = t$ for $t \in [0, 1]$ ($Y \sim U(0, 1)$). This fact has an important implication: if the task is to generate a sample from the distribution with a CDF F that admits the inverse F^{-1} and there is a random sample generator for the uniform probability distribution $U(0, 1)$, then the sample can be obtained by applying F^{-1} to a sample drawn from $U(0, 1)$. For example, a sample of the normal distribution with mean μ and standard deviation σ can be generated by applying the inverse function of $\Phi\left(\frac{t-\mu}{\sigma}\right)$ to a sample from $U(0, 1)$, where $\Phi(t) = \frac{1}{\sqrt{2\pi}} \int_{-\infty}^{t} e^{-s^2/2} ds$ is the CDF of the standard normal random variable (with $\mu = 0$ and $\sigma = 1$).

Similarly, a function $g : \mathbb{R}^n \to \mathbb{R}$ of an n-dimensional random vector $X = (X_1, \ldots, X_n)$ is a random variable $Y = g(X)$ with the CDF $F_Y(t) = \mathbb{P}[Y \le t] = \mathbb{P}[g(X) \le t]$. If $X_1, \ldots, X_n$ are continuously distributed with a joint PDF $f_X(t_1, \ldots, t_n)$, then $F_Y(t)$ is determined by the n-dimensional integral of f_X over the region $\{s = (s_1, \ldots, s_n) \in \mathbb{R}^n | g(s) \le t\}$:

$$F_Y(t) = \int \cdots \int_{\{s=(s_1,\ldots,s_n)\in\mathbb{R}^n | g(s)\le t\}} f_X(s_1, \ldots, s_n)\, ds_1 \ldots ds_n.$$

The structural dependence of random variables $X_1, \ldots, X_n$ is completely determined by their joint CDF $F_X(t_1, \ldots, t_n)$. Often, it is also characterized through *copula*. Let $F_{X_k}(t)$ be the CDF of X_k, $k = 1, \ldots, n$. Then the random variable $Y_k = F_{X_k}(X_k)$, $k \in \{1, \ldots, n\}$, has the uniform distribution $U(0, 1)$, and the joint CDF of $Y_1, \ldots, Y_n$ is called copula denoted by $C(t_1, \ldots, t_n)$. In other words, $C(t_1, \ldots, t_n) = \mathbb{P}[Y_1 \leq t_1, \ldots, Y_n \leq t_n]$. If all $F_{X_1}(t), \ldots, F_{X_n}(t)$ admit inverse, then $F_X(t_1, \ldots, t_n) = C(F_{X_1}^{-1}(t_1), \ldots, F_{X_n}^{-1}(t_n))$. For example, a sample of a random vector $X = (X_1, \ldots, X_n)$ with continuous marginal CDFs $F_{X_1}(t), \ldots, F_{X_n}(t)$ can be obtained as follows: generate a sample of the random vector $Y = (Y_1, \ldots, Y_n)$ having the CDF $C(t_1, \ldots, t_n)$ and then apply $F_{X_k}^{-1}$ to the kth component of each vector in the generated sample. If $X_1, \ldots, X_n$ have the multivariate normal distribution with mean $\mu = (\mu_1, \ldots, \mu_n)$ and covariance matrix Λ, then $C(t_1, \ldots, t_n) = \mathbb{P}[\Phi_1(X_1) \leq t_1, \ldots, \Phi_n(X_n) \leq t_n]$ is called *Gaussian copula*, where $\Phi_k(t)$ is the marginal normal distribution of X_k. Similarly, the Gaussian copula is used to generate samples of a random vector, whose components are multivariate normally distributed with given marginals.

1.2 Moments and Moment-Generating Function

The nth-order moment and nth-order central moment of a random variable X with the CDF $F_X(t)$ are defined by

$$E\left[X^n\right] = \int_{-\infty}^{\infty} t^n dF_X(t)$$

and

$$E\left[(X - E[X])^n\right] = \int_{-\infty}^{\infty} (t - E[X])^n \, dF_X(t),$$

respectively, where $E[\cdot]$ is the *expectation* operator. They are fundamental integral characteristics of X. For example, the mean $\mu(X)$ and variance $\sigma^2(X)$ are the first moment and the second central moment of X, respectively:

$$\mu(X) = E[X], \qquad \sigma^2(X) = E\left[(X - E[X])^2\right].$$

Also, *skewness* is a measure of the asymmetry of $F_X(t)$ defined through the third central moment

$$\frac{E\left[(X - E[X])^3\right]}{\sigma^3(X)},$$

whereas *kurtosis* is a measure of tail fatness defined via the fourth central moment

$$\frac{E\left[(X - E[X])^4\right]}{\sigma^4(X)},$$

provided that the third and fourth moments are finite, respectively.

The pth-order absolute moment $E\big[|X|^p\big]$, $p \in [1,\infty)$, is closely related to p-norm $\|\cdot\|_p$ in the Lebesgue space $\mathscr{L}^p(\Omega)$. Namely, $\|\cdot\|_p$ is defined by

$$\|X\|_p = \begin{cases} \big(E\big[|X|^p\big]\big)^{1/p} & \text{for } p \in [1,\infty), \\ \sup |X| & \text{for } p = \infty, \end{cases}$$

and $\mathscr{L}^p(\Omega) = \big\{X \mid \|X\|_p < \infty\big\}$. If X represents a random error in some approximation or measurement, then $\|X\|_p$ can be used as an error measure of X.

The first-order absolute central moment $E[|X - E[X]|]$ is known as *mean absolute deviation* (MAD) of X:

$$\mathrm{MAD}(X) = E[|X - E[X]|].$$

While absolute moments are instrumental for constructing norms and error measures, absolute central moments such as MAD and variance give rise to the class of deviation measures called "penalties relative to expectation." The theory of deviation, risk, and error measures is discussed in Chap. 2.

1.2.1 Classical Inequalities with Moments

We proceed with stating classical inequalities involving moments:

- *Lyapunov's inequality* for absolute moments of a random variable $X \in \mathscr{L}^q(\Omega)$:

$$E\left[|X|^p\right]^{\frac{1}{p}} \le E\left[|X|^q\right]^{\frac{1}{q}} \qquad \text{for } 0 < p < q,$$

  or equivalently, $\|X\|_p \le \|X\|_q$ for $0 < p < q$, which implies that

$$\|X\|_1 \le \|X\|_2 \le \|X\|_3 \le \dots$$

- *Schwarz's inequality*: for two random variables $X \in \mathscr{L}^2(\Omega)$ and $Y \in \mathscr{L}^2(\Omega)$, we have

$$(E[X\,Y])^2 \equiv \left(\int_\Omega X(\omega)\,Y(\omega)\,d\mathbb{P}[\omega]\right)^2 \le E\left[X^2\right]E\left[Y^2\right].$$

  If in Schwarz's inequality, X and Y are replaced by $(X - E[X])/\sigma(X)$ and $(Y - E[Y])/\sigma(Y)$, respectively, then we obtain a well-known fact that

$$\mathrm{Cov}(X,Y) \equiv E[(X - E[X])(Y - E[Y])] \le \sigma(X)\,\sigma(Y).$$

- *Jensen's inequality*: if f is a convex function[2] on the range of a random variable X, then

$$E[f(X)] \geq f(E[X]).$$

As its immediate corollary, we obtain that if f is a concave function on the range of X, then

$$E[f(X)] \leq f(E[X]).$$

(It is enough to observe that in this case, $-f$ is convex.) Examples of Jensen's inequality include

$$E\left[\mathrm{e}^X\right] \geq \mathrm{e}^{E[X]}, \qquad E\left[\ln X\right] \leq \ln(E[X]) \quad \text{for } X > 0,$$

and

$$|E[X]|^p \leq (E[|X|])^p \leq E[|X|^p] \qquad \text{for } p > 1.$$

The latter also follows from Lyapunov's inequality.

1.2.2 Moment-Generating Function

The *moment-generating function* (MGF) of a random variable X is defined by

$$M_X(t) = E\left[\mathrm{e}^{tX}\right], \qquad t \in \mathbb{R}.$$

If $M_X(t)$ exists in some neighborhood of $t = 0$, then the nth moment of X is the nth-order derivative of $M_X(t)$ at $t = 0$:

$$E\left[X^n\right] = \left.\frac{d^n}{dt^n} E\left[\mathrm{e}^{tX}\right]\right|_{t=0}.$$

For example, if a random variable X is distributed uniformly on the interval $[a, b]$, then $M_X(t) = \left(\mathrm{e}^{bt} - \mathrm{e}^{at}\right)/((b-a)t)$ with $\lim_{t\to 0} M_X(t) = 1$, whereas if it has the normal distribution with mean μ and standard deviation σ, then $M_X(t) = \mathrm{e}^{\mu t + \sigma^2 t^2/2}$.

[2]A function $f(x) : [a, b] \to \mathbb{R}$ is convex on $[a, b]$ with $a < b$ if $f(\lambda x_1 + (1-\lambda)x_2) \leq \lambda f(x_1) + (1-\lambda) f(x_2)$ for any $x_1 \in [a, b]$ and $x_2 \in [a, b]$ and any $\lambda \in [0, 1]$. This definition also holds if the domain of $f(x)$ is $\mathbb{R}$, i.e., for $f : \mathbb{R} \to \mathbb{R}$. A function $g(x)$ is concave on a given interval if $-g(x)$ is convex on that interval.

The moment-generating function has the following properties provided that it exists in some neighborhood of $t = 0$:

- All moments of a random variable X are finite, and

$$M_X(t) = \sum_{n=0}^{\infty} \frac{E[X^n]}{n!} t^n.$$

In general, X may have all finite moments, but $M_X(t)$ may not exist for $t \neq 0$.
- If two random variables X and Y have the same moment-generating function, i.e., $M_X(t) = M_Y(t)$, then X and Y have the same CDF: $F_X(t) = F_Y(t)$.
- If random variables $X_1, \ldots, X_n$ are independent with moment-generating functions $M_1(t), \ldots, M_n(t)$, respectively, then

$$M_{X_1+\cdots+X_n}(t) = \prod_{k=1}^{n} M_k(t).$$

1.3 Partial Moments and Stochastic Dominance

The nth-order lower and upper partial moments of a random variable X, also known as one-sided moments, are defined with respect to a reference point c as

$$E[([c - X]_+)^n] \qquad \text{and} \qquad E[([X - c]_+)^n],$$

respectively, where $[t - c]_+ = \max\{t - c, 0\}$. They are used to characterize the left and right parts (tails) of the distribution of X. For example, if $c = \mu = E[X]$, then

$$E[[\mu - X]_+] = E[[X - \mu]_+] = \frac{1}{2}\mathrm{MAD}(X),$$

whereas

$$\sigma_-(X) = \sqrt{E[([\mu - X]_+)^2]}, \qquad \sigma_+(X) = \sqrt{E[([X - \mu]_+)^2]}$$

are called *standard lower* and *standard upper semideviations* of X, respectively, and

$$\sigma_-^2(X) + \sigma_+^2(X) = \sigma^2(X).$$

If X represents the random return of a portfolio of financial assets, then $E[[\mu - X]_+]$ and $\sigma_-(X)$ measure the extent of portfolio underperformance with respect to the portfolio expected return and are desired to be as small as possible.

1.3.1 Performance Functions

Nested integrals of the CDF $F_X(t)$ of a random variable X are defined by

$$F_X^{(1)}(t) = F_X(t) \qquad \text{and} \qquad F_X^{(n)}(t) = \int_{-\infty}^{t} F_X^{(n-1)}(s)\, ds \quad \text{for } n \ge 2$$

and are also called performance functions.

If $E[|X|^n] < \infty$, then the nth-order lower partial moment can be expressed as

$$E[([c - X]_+)^n] = \int_{-\infty}^{c} (c - t)^n dF_X(t) = n!\, F_X^{(n+1)}(c);$$

see [33] for the proof. As immediate corollaries of this relationship, we have

$$\mathrm{MAD}(X) = 2F_X^{(2)}(\mu), \qquad \mu = E[X], \tag{1.3.1}$$

provided that $E[|X|] < \infty$, and

$$\sigma_-^2(X) = 2F_X^{(3)}(\mu), \qquad \mu = E[X], \tag{1.3.2}$$

provided that $E\left[X^2\right] < \infty$.

1.3.2 Stochastic Dominance

Stochastic dominance, also called *stochastic ordering*, is one of the fundamental notions in decision analysis, which originates from the expected utility theory and introduces a preference relation $\succeq$ between two random variables X and Y to hold for all von Neumann–Morgenstern utility functions from a certain class. For example, the *first-order* stochastic dominance (FSD) relation $X \succeq_{(1)} Y$ means that $E[U(X)] \ge E[U(Y)]$ should hold for all *nondecreasing* utility functions U, whereas the *second-order* stochastic dominance (SSD) relation $X \succeq_{(2)} Y$ implies that $E[U(X)] \ge E[U(Y)]$ holds for all *nondecreasing* and *concave* utility functions U. Stochastic dominance can be conveniently characterized in terms of performance functions. Namely, a random variable X dominates a random variable Y with respect to nth-order stochastic dominance if and only if $F_X^{(n)}(t) \le F_Y^{(n)}(t)$ for all $t \in \mathbb{R}$, where $F_X^{(n)}(t)$ and $F_Y^{(n)}(t)$ are performance functions of X and Y, respectively, and we write

$$X \succeq_{(n)} Y \qquad \Longleftrightarrow \qquad F_X^{(n)}(t) \le F_Y^{(n)}(t) \quad \forall t \in \mathbb{R}.$$

In particular, $X \succeq_{(n)} Y$ implies $X \succeq_{(n+1)} Y$ for $n \in \mathbb{N}$. Also, if $E[|X|^n] < \infty$ and $E[|Y|^n] < \infty$ for some $n \in \mathbb{N}$, and $X \succeq_{(n+1)} Y$, then $E[X] \geq E[Y]$. For $n = 1$, this fact is obvious, since the SSD implies that $E[U(X)] \geq E[U(Y)]$ for all nondecreasing and concave utility functions U including $U(X) \equiv X$.

The SSD is often associated with *risk-averse* decision making.[3] It elucidates the role of standard deviation, MAD, and standard lower semideviation in risk analysis. Namely, if $E[X] = E[Y]$ and $X \succeq_{(2)} Y$, then

- $\sigma(X) \leq \sigma(Y)$; see the proof in [28].
- (1.3.1) implies that $\mathrm{MAD}(X) \leq \mathrm{MAD}(Y)$.
- (1.3.2) implies that $\sigma_-(X) \leq \sigma_-(Y)$ since $X \succeq_{(3)} Y$ follows from $X \succeq_{(2)} Y$.

In other words, if a risk-averse decision maker uses any of these three measures (σ, MAD, or σ_-) to order random variables with equal means, then he/she always prefers the one with the least corresponding measure. This is one of the key arguments for Markowitz's mean-variance approach.

1.4 Quantile Functions and Fenchel Duality

Let X be an arbitrary random variable, and let $F_X(t)$ be the CDF of X. Then for $s \in (0, 1)$, an s-quantile of X is any value in the interval $[q_X^-(s), q_X^+(s)]$, where

$$q_X^-(s) = \inf\{t \mid F_X(t) \geq s\} \quad \text{and} \quad q_X^+(s) = \sup\{t \mid F_X(t) \leq s\}$$

are called *lower* and *upper s-quantiles*, respectively, and are related by

$$q_{-X}^+(s) = -q_X^-(1-s).$$

Observe that $q_X^+(s) = q_X^-(s) = F_X^{(-1)}(s)$ for a continuously distributed X. Often, either $q_X^+(s)$ or $q_X^-(s)$ is considered as a generalized inverse of $F_X(t)$.[4] Also, when X has no probability "atom" at $q_X^-(s)$, i.e., when $F_X(t)$ is continuous at $t = q_X^-(s)$, the interval $\left[q_X^-(s), q_X^+(s)\right]$ collapses to a single value.

Hardy–Littlewood's inequality, extended for random variables X and Y, yields

$$E[XY] \equiv \int_\Omega X(\omega)\, Y(\omega)\, d\mathbb{P}[\omega] \leq \int_0^1 q_X^-(s)\, q_Y^-(s)\, ds,$$

provided that both sides of the inequality exist; see [18] and [13, Theorem A.24].

---

[3]Given a choice between a nonconstant random variable X and its expected value $E[X]$, a risk-averse agent always prefers $E[X]$ over X.

[4]In fact, $F_X^{(-1)}(s)$ is often defined as $q_X^-(s)$; see, e.g., [34].

Similarly to the nested integrals of $F_X(t)$ of a random variable X, we can define

$$q_X^{(1)}(s) = q_X^+(s) \qquad \text{and} \qquad q_X^{(n)}(s) = \int_0^s q_X^{(n-1)}(\tau)\, d\tau \quad \text{for } n \geq 2.$$

For example, $q_X^{(2)}(s)$ is closely related to the *average s-quantile* $\overline{q}_X(s)$:

$$\overline{q}_X(s) \equiv \frac{1}{s} \int_0^s q_X^+(\tau)\, d\tau = \frac{1}{s} q_X^{(2)}(s). \tag{1.4.1}$$

Observe that $\overline{q}_X(s) \to \inf X$ as $s \to 0$ and $\overline{q}_X(s) \to E[X]$ as $s \to 1$. By continuity, we define $\overline{q}_X(0) = \inf X$ and $\overline{q}_X(1) = E[X]$.

Example 1.1 (average s-quantile). Let X be a discretely distributed random variable with the probability distribution $\mathbb{P}[X = x_k] = p_k > 0$, $k = 1, \ldots, n$, where $\sum_{k=1}^n p_k = 1$ and $x_1 < \cdots < x_n$. In this case,

$$q_X^+(s) = \begin{cases} x_1, & s \in [0, p_1), \\ x_2, & s \in [p_1, p_1 + p_2), \\ \ldots & \ldots \\ x_k, & s \in [p_1 + \cdots + p_{k-1}, p_1 + \cdots + p_k), \\ \ldots & \ldots \\ x_n, & s \in [p_1 + \cdots + p_{n-1}, 1), \end{cases}$$

and the formula (1.4.1) yields

$$\overline{q}_X(s) = \begin{cases} x_1 \text{ if } s \in [0, p_1), \\ \dfrac{p_1 x_1 + \cdots + p_k x_k + (s - p_1 - \cdots - p_k) x_{k+1}}{s} \\ \qquad \text{if } s \in [p_1 + \cdots + p_k, p_1 + \cdots + p_{k+1}) \text{ for } k = 1, \ldots, n-1, \\ p_1 x_1 + \cdots + p_n x_n = E[X] \text{ if } s = 1. \end{cases}$$

In particular, if $s = p_1 + \cdots + p_m$ for some $m \in \{1, \ldots, n\}$, then $\overline{q}_X(s) = (p_1 x_1 + \cdots + p_m x_m)/(p_1 + \cdots + p_m)$.

Negative $\overline{q}_X(s)$ is also called *conditional value-at-risk (CVaR)* [39, 40] and *expected shortfall (ES)* [1] discussed in Chap. 2. As a corollary from (1.4.1), we can write

$$q_{-X}^{(2)}(s) = q_X^{(2)}(1 - s) - E[X],$$

$$\overline{q}_{-X}(s) = \frac{1}{s} \left((1 - s)\, \overline{q}_X(1 - s) - E[X] \right).$$

Remarkably, $F_X^{(2)}(t)$ and $q_X^{(2)}(s)$ are related through *Fenchel duality*, namely

$$q_X^{(2)}(s) = \sup_{t\in\mathbb{R}} \left(s\,t - F_X^{(2)}(t) \right) \tag{1.4.2}$$

and

$$F_X^{(2)}(t) = \sup_{s\in(0,1)} \left(s\,t - q_X^{(2)}(s) \right); \tag{1.4.3}$$

see, e.g., [33, 34] for the proof.

In (1.4.2), the supremum is attained at any value from the quantile interval $[q_X^-(s), q_X^+(s)]$, so that

$$\begin{aligned} q_X^{(2)}(s) &= s\,q_X^+(s) - F_X^{(2)}(q_X^+(s)) \equiv s\,q_X^+(s) - E[[q_X^+(s) - X]_+] \\ &= s\,q_X^-(s) - F_X^{(2)}(q_X^-(s)) \equiv s\,q_X^-(s) - E[[q_X^-(s) - X]_+], \end{aligned}$$

whereas in (1.4.3), the supremum is attained at any $s \in [F_X(t-), F_X(t)]$. These relationships and (1.4.1) imply

$$\begin{aligned} \overline{q}_X(s) &= q_X^+(s) - \frac{1}{s} E[[q_X^+(s) - X]_+] \\ &= q_X^-(s) - \frac{1}{s} E[[q_X^-(s) - X]_+]. \end{aligned}$$

Definition (1.4.1) and representation (1.4.2) yield the optimization formula for the average quantile; see, e.g., [39, 40]:

$$\overline{q}_X(s) = \max_{c\in\mathbb{R}} \left(c - \frac{1}{s} E[[c - X]_+] \right), \tag{1.4.4}$$

in which optimal c is any value from the quantile interval $[q_X^-(s), q_X^+(s)]$.

The significance of the formula (1.4.4) *is that in the case of a discretely distributed X, it reduces the tedious task of calculating the average of $s \cdot 100\%$ of the worst (lowest) outcomes of X to a simple optimization procedure.* Indeed, suppose X is discretely distributed with the probability distribution $\mathbb{P}[X = x_k] = p_k > 0$, $k = 1, \ldots, n$, where $\sum_{k=1}^n p_k = 1$. Then (1.4.4) takes the form of a linear program

$$\begin{aligned} \overline{q}_X(s) = &\max_{c,\, y_1,\ldots,y_n} \left(c - \frac{1}{s} \sum_{k=1}^{n} p_k\, y_k \right) \\ &\text{subject to} \quad y_k \ge c - x_k, \quad y_k \ge 0, \quad k = 1, \ldots, n, \end{aligned} \tag{1.4.5}$$

where $y_1, \ldots, y_n$ are auxiliary variables. The dual problem for (1.4.5) is given by

$$\overline{q}_X(s) = \min_{z_1,\ldots,z_n} \sum_{k=1}^{n} z_k p_k x_k$$
$$\text{subject to} \quad \sum_{k=1}^{n} z_k p_k = 1, \quad 0 \leq z_k \leq 1/s, \quad k = 1, \ldots, n. \tag{1.4.6}$$

The FSD and SSD can also be characterized through the quantile functions:

$$X \succeq_{(1)} Y \quad \Longleftrightarrow \quad q_X^-(s) \geq q_Y^-(s) \quad \text{for all } s \in (0,1),$$

and

$$X \succeq_{(2)} Y \quad \Longleftrightarrow \quad q_X^{(2)}(s) \geq q_Y^{(2)}(s) \quad \text{for all } s \in (0,1).$$

1.5 Entropy and Distribution Divergence Measures

1.5.1 Shannon Entropy and Kullback–Leibler Divergence

Shannon entropy is one of the fundamental notions in information theory and statistics and is closely related to the *Gibbs entropy* in thermodynamics. It is widely used as a measure of uncertainty in a random variable X. The higher the entropy of X is, the more X is considered to be uncertain. For example, among random variables distributed on $[a,b]$, the one with the uniform distribution has the highest entropy, whereas a constant has the lowest entropy, which vanishes on a discrete probability space.

For a discretely distributed random variable X assuming n distinct values $x_1, \ldots, x_n$ with positive probabilities $p_1, \ldots, p_n$, the Shannon entropy is defined by

$$S(X) = -\sum_{k=1}^{n} p_k \ln p_k, \tag{1.5.1}$$

whereas for a continuously distributed random variable X with the PDF $f_X(t) > 0$ on $[a,b]$, the Shannon entropy, also known as *Shannon differential entropy*, is given by

$$S(X) = -\int_a^b f_X(t) \ln f_X(t) dt. \tag{1.5.2}$$

The Shannon entropy (1.5.1) attains its maximum $\ln n$ at $p_1 = \cdots = p_n = 1/n$ and is always nonnegative, whereas the Shannon differential entropy (1.5.2) attains its maximum $\ln(b-a)$ at $f_X(t) = 1/(b-a)$, $t \in [a,b]$, and can be negative, for example when $b - a < 1$.

Remark 1.1. Since $\lim_{\epsilon \to 0+} \epsilon \ln \epsilon = 0$, definitions (1.5.1) and (1.5.2) hold for $p_1 \geq 0, \ldots, p_n \geq 0$ and for $f_X(t) \geq 0$ on $[a,b]$, respectively, i.e., when $p_k = 0$ for some $k \in \{1, \ldots, n\}$ and when $f_X(t) = 0$ for some $t \in [a,b]$.

The notion of entropy gives rise to various *entropic* functionals for measuring closeness of random variables in the sense of their probability distributions, which are known as *probability divergence measures.*

If X and Y are two discretely distributed random variables defined on the same probability space of elementary events with the probability distributions $\mathbb{P}[X = x_k] = p_k \geq 0$ and $\mathbb{P}[Y = y_k] = q_k > 0$, $k = 1, \ldots, n$, respectively, then the *Kullback–Leibler divergence measure* of X and Y, also called *relative entropy,* measures the closeness of the distributions of X and Y and is defined by

$$D_{\mathrm{KL}}(X||Y) = \sum_{k=1}^{n} p_k \ln \frac{p_k}{q_k}. \tag{1.5.3}$$

Note that $D_{\mathrm{KL}}(X||Y)$ is asymmetric with respect to X and Y: $D_{\mathrm{KL}}(X||Y) \neq D_{\mathrm{KL}}(Y||X)$.

Since $-\ln t$ is a convex function, Jensen's inequality in the discrete form[5] yields

$$D_{\mathrm{KL}}(X||Y) = -\sum_{k:\, p_k>0} p_k \ln \frac{q_k}{p_k} \geq -\ln\left(\sum_{k:\, p_k>0} p_k \frac{q_k}{p_k}\right) \geq -\ln 1 = 0,$$

so that the Kullback–Leibler divergence measure is always nonnegative:

$$D_{\mathrm{KL}}(X||Y) \geq 0,$$

where the equality is attained when X and Y have the same probability distribution. A quick corollary from this fact is the *Gibbs' inequality*

$$-\sum_{k=1}^{n} p_k \ln q_k \geq -\sum_{k=1}^{n} p_k \ln p_k = S(X).$$

[5]If $f : \mathbb{R} \to \mathbb{R}$ is a convex function, then $f(\sum_{k=1}^{n} \lambda_k t_k) \leq \sum_{k=1}^{n} \lambda_k f(t_k)$ for any $t_1, \ldots, t_n$ and $\lambda_1 \geq 0, \ldots, \lambda_n \geq 0$ such that $\sum_{k=1}^{n} \lambda_k = 1$. This follows from the fact that the set $\{(t,s) \in \mathbb{R} \times \mathbb{R} \mid f(t) \leq s\}$, called *epigraph* of f, is convex; see [27, Proposition 6.1.1] and [51, Lemma 2.60].

Similarly, for continuously distributed random variables X and Y on $[a,b]$ with PDFs $f_X(t) \geq 0$ and $f_Y(t) > 0$, respectively, the Kullback–Leibler divergence measure is defined by

$$D_{\mathrm{KL}}(X||Y) = \int_a^b f_X(t) \ln \frac{f_X(t)}{f_Y(t)} dt,$$

which is nonenegative by Jensen's integral inequality [23]. Indeed, let $\mathscr{T} = \{t \in [a,b] \mid f_X(t) > 0\}$, then

$$\begin{aligned} D_{\mathrm{KL}}(X||Y) &= \int_{\mathscr{T}} f_X(t) \left(-\ln \frac{f_Y(t)}{f_X(t)} \right) dt \geq -\ln \left(\int_{\mathscr{T}} f_X(t) \frac{f_Y(t)}{f_X(t)} dt \right) \\ &\geq -\ln 1 = 0. \end{aligned}$$

In Bayesian statistics, the Kullback–Leibler divergence measure is interpreted as the information gain of a posterior distribution over a prior distribution and is instrumental in statistical hypothesis testing. Suppose a random variable X has different distributions under two mutually exclusive hypotheses H_1 and H_0. For clarity, let $X = X_1$ and $X = X_0$ under H_1 and H_0, respectively. Then $D_{\mathrm{KL}}(X_1||X_0)$ is considered to be the expected discrimination information for H_1 over H_0.

Minimization of the Kullback–Leibler divergence measure is also known as *relative entropy minimization* and the *principle of minimum discrimination information*.

Let $\lambda X \oplus (1-\lambda)Y$ with $\lambda \in (0,1)$ denote a random variable Z with the CDF F_Z given by the distributional mixture of the CDFs F_X and F_Y: $F_Z = \lambda F_X + (1-\lambda) F_Y$. In the finance literature, Z is also called *lottery* between X and Y, which realize with probabilities λ and $1-\lambda$, respectively. Then λ-*divergence* is determined by

$$D_\lambda(X||Y) = \lambda D_{\mathrm{KL}}(X \,||\, \lambda X \oplus (1-\lambda)Y) + (1-\lambda) D_{\mathrm{KL}}(Y \,||\, \lambda X \oplus (1-\lambda)Y).$$

It is often used in various probability distribution estimation problems.

1.5.2 Renyi Entropy and Renyi Divergence

The *Renyi entropy* is arguably one of the most well-known generalizations of the Shannon entropy. For a discretely distributed random variable X with the distribution $\mathbb{P}[X = x_k] = p_k > 0$, $k = 1, \ldots, n$, the Renyi entropy of order $\alpha \geq 0$ is defined by

$$H_\alpha(X) = \begin{cases} \dfrac{1}{1-\alpha} \ln \left(\sum_{k=1}^{n} p_k^\alpha \right), & \alpha \neq 1, \\ -\sum_{k=1}^{n} p_k \ln p_k, & \alpha = 1. \end{cases} \tag{1.5.4}$$

It has several special cases:

(a) $H_0(X) = \ln n$ is also known as the *Hartley entropy*.
(b) $H_1(X)$ coincides with the Shannon entropy.
(c) $H_2(X)$ is known as *collision entropy*.
(d) $H_\infty(X) = \min_{k=1,\ldots,n} (-\ln p_k) \equiv -\ln \left(\max_{k=1,\ldots,n} p_k \right)$ is called *min-entropy*.

For a continuously distributed random variable X with the PDF $f_X(t)$ on support $[a, b]$, the *Renyi differential entropy* of order $\alpha > 0$ has a similar definition:

$$H_\alpha(X) = \begin{cases} \dfrac{1}{1-\alpha} \ln \left(\int_a^b f_X(t)^\alpha dt \right), & \alpha \neq 1, \\ -\int_a^b f_X(t) \ln f_X(t)\, dt, & \alpha = 1. \end{cases} \tag{1.5.5}$$

The *Renyi divergence* of order $\alpha > 0$ is a generalization of the Kullback–Leibler divergence (relative entropy). For two random variables X and Y with probability distributions $\mathbb{P}[X = x_k] = p_k \geq 0$ and $\mathbb{P}[Y = y_k] = q_k > 0$, $k = 1, \ldots, n$, respectively, it is defined by

$$D_\alpha(X||Y) = \begin{cases} \dfrac{1}{\alpha-1} \ln \left(\sum_{k=1}^{n} p_k^\alpha q_k^{1-\alpha} \right), & \alpha \neq 1, \\ \sum_{k=1}^{n} p_k \ln \dfrac{p_k}{q_k}, & \alpha = 1, \end{cases} \tag{1.5.6}$$

whereas for continuously distributed X and Y on $[a, b]$ with PDFs $f_X(t) \geq 0$ and $f_Y(t) > 0$, respectively, it is defined by

$$H_\alpha(X) = \begin{cases} \dfrac{1}{\alpha-1} \ln \left(\int_a^b f_X(t)^\alpha f_Y(t)^{1-\alpha} dt \right), & \alpha \neq 1, \\ \int_a^b f_X(t) \ln \dfrac{f_X(t)}{f_Y(t)}\, dt, & \alpha = 1. \end{cases} \tag{1.5.7}$$

As the Shannon entropy and Kullback–Leibler divergence, the Renyi entropy and Renyi divergence are used in estimating unknown probability distributions with the maximum entropy and minimum relative entropy principles as well as in testing of statistical hypotheses.

Chapter 2
Deviation, Risk, and Error Measures

In statistical decision theory, uncertainty inherent in a random variable X is often translated into although related but not the same notions of *error*, *deviation* (*dispersion*), and *risk* revolving around the standard deviation $\sigma(X)$. By definition, $\sigma(X)$ is a measure of *deviation* of X from its expected value $E[X]$, i.e., $\sigma(X) = \|X - E[X]\|_2$. It is closely related to measurement of uncertainty in outcomes, i.e., to *dispersion*, to aggregated measurement of probable undesirable outcomes (losses), i.e., to *risk*, and to measurement of quality of estimation in statistics, i.e., to *error*. For example, in classical portfolio theory [29], variance, or equivalently σ, is used to quantify uncertainty in returns of financial portfolios, whereas the difference between the standard deviation and the expected value of portfolio return can be interpreted as risk. Therefore, with the standard deviation, we may associate a triplet $\langle \mathscr{D}, \mathscr{R}, \mathscr{E} \rangle$: deviation measure $\mathscr{D}(X) = \sigma(X) \equiv \|X - E[X]\|_2$, risk measure $\mathscr{R}(X) = \sigma(X) - E[X] \equiv \|X - E[X]\|_2 - E[X]$, and error measure $\mathscr{E}(X) = \|X\|_2$.

Another well-known example of such a triplet is the one associated with the mean absolute deviation (MAD), which sometimes is used instead of the standard deviation. In this case, $\mathscr{D}$, $\mathscr{R}$, and $\mathscr{E}$ are defined by $\mathscr{D}(X) = \|X - E[X]\|_1$, $\mathscr{R}(X) = \|X - E[X]\|_1 - E[X]$, and $\mathscr{E}(X) = \|X\|_1$. Obviously, the triplet $\mathscr{D}(X) = \|X - E[X]\|_p$, $\mathscr{R}(X) = \|X - E[X]\|_p - E[X]$, and $\mathscr{E}(X) = \|X\|_p$ with $p \geq 1$ generalizes the previous two. However, none of these standard triplets are appropriate for applications involving noticeably asymmetric distributions of outcomes.

In financial applications, *value-at-risk (VaR)*, defined as the negative upper α-quantile

$$\mathrm{VaR}_\alpha(X) = -q_X^+(\alpha) \tag{2.0.1}$$

emerged as a major competitor to the standard deviation and MAD. However, as a measure of risk, $\mathrm{VaR}_\alpha(X)$ lacks convexity and provides no information of how significant losses in the α-tail could be. These VaR's deficiencies are resolved by *conditional value-at-risk (CVaR)* [39, 40] defined as the average of VaR on $[0, \alpha]$:

M. Zabarankin and S. Uryasev, *Statistical Decision Problems: Selected Concepts and Portfolio Safeguard Case Studies*, Springer Optimization and Its Applications 85,
DOI 10.1007/978-1-4614-8471-4_2, © Springer Science+Business Media New York 2014

$$\mathrm{CVaR}_\alpha(X) = \frac{1}{\alpha}\int_0^\alpha \mathrm{VaR}_s(X)\,ds, \tag{2.0.2}$$

which is equivalent to the negative average α-quantile of X, i.e., $-\overline{q}_X(\alpha)$;[1] see Example 1.1 and the formula (1.4.4) for evaluating $\overline{q}_X(\alpha)$. Similarly to the standard deviation and MAD, CVaR induces the triplet: CVaR deviation $\mathscr{D}_\alpha(X) = \mathrm{CVaR}_\alpha(X - E[X])$, CVaR-risk measure $\mathscr{R}_\alpha(X) = \mathrm{CVaR}_\alpha(X)$, and *asymmetric mean absolute error*

$$\mathscr{E}_\alpha(X) = \frac{1}{\alpha}E[\alpha\, X_+ + (1-\alpha)\, X_-], \qquad \alpha \in (0,1), \tag{2.0.3}$$

where $X_\pm = \max\{0, \pm X\}$. For $\alpha = 1/2$, $\mathscr{E}_\alpha(X)$ simplifies to $\|X\|_1$.

Practical needs motivated a search for other triplets which could preserve consistency in risk preferences and would provide adequate analysis of asymmetric distributions in related decision problems. For example, if an agent uses standard lower semideviation σ_- in a portfolio selection problem, it is expected that the agent would use a corresponding error measure in an asset pricing factor model. In response to these needs, Rockafellar et al. [41, 42, 46] developed a coordinating theory of *deviation measures*, *error measures*, and *averse measures of risk*, which, in general, are not symmetric with respect to ups and downs of X. Deviation measures [42] quantify *nonconstancy* in X and preserve four main properties of the standard deviation (*nonnegativity*, *positive homogeneity*, *subadditivity*, and *insensitivity to constant shift*), whereas *error measures* [46] quantify *nonzeroness* of X and generalize the expected square error. The triplets $\langle \mathscr{D}, \mathscr{R}, \mathscr{E}\rangle$ for the standard deviation, MAD and CVaR are, in fact, particular examples of more general relationships

$$\mathscr{R}(X) = \mathscr{D}(X) - E[X],$$
$$\mathscr{D}(X) = \min_{c\in\mathbb{R}} \mathscr{E}(X-c), \quad \text{or} \quad \mathscr{D}(X) = \mathscr{E}(X - E[X]).$$

In this theory, risk, deviation, and error measures are *lower semicontinuous positively homogenous convex* functionals, satisfying closely related systems of axioms. They were comprehensively studied in the framework of convex analysis [36, 37]. Rockafellar et al. [43–46] generalized the mean-variance approach to portfolio selection and derived optimality conditions for a linear regression with general error measures, whereas Grechuk et al. [16, 17] extended Chebyshev's inequality and the maximum entropy principle for *law-invariant* deviation measures (i.e., those that depend only on the distribution of X).

---

[1]In finance, $L = -X$ is often interpreted as loss, and *CVaR for loss*, denoted by CVaR_α^-, is defined as the average of the right $(1-\alpha)$-tail of L, i.e., $\mathrm{CVaR}_\alpha^-(L) = \frac{1}{1-\alpha}\int_\alpha^1 q_L^-(s)\,ds$, where $q_L^-(s)$ is the lower s-quantile of L; see [39, 40]. In fact, $\mathrm{CVaR}_\alpha^-(L) = \mathrm{CVaR}_{1-\alpha}(X)$.

2.1 Deviation Measures

Responding to the need for flexibility in treating the ups and downs of a random outcome differently, Rockafellar et al. [42] introduced a deviation measure as a functional $\mathscr{D} : \mathscr{L}^2(\Omega) \to [0, \infty]$ satisfying the axioms

(D1) *Nonnegativity*: $\mathscr{D}(X) = 0$ for constant X, but $\mathscr{D}(X) > 0$ otherwise.
(D2) *Positive homogeneity*: $\mathscr{D}(\lambda X) = \lambda \mathscr{D}(X)$ when $\lambda > 0$.
(D3) *Subadditivity*: $\mathscr{D}(X + Y) \leq \mathscr{D}(X) + \mathscr{D}(Y)$ for all X and Y.
(D4) *Lower semicontinuity*: set $\{X \in \mathscr{L}^2(\Omega) | \mathscr{D}(X) \leq c\}$ is closed for all $c < \infty$.

It follows from D1 and D3 that

$$\mathscr{D}(X - c) = \mathscr{D}(X) \qquad \text{for all constants } c,$$

which is known as *insensitivity to constant shift* (see [42]).

Axioms D1–D4 generalize well-known properties of the standard deviation; however, they do not imply symmetry, so that in general, $\mathscr{D}(-X) \neq \mathscr{D}(X)$. A deviation measure is called *lower range dominated* if, in addition to D1–D4, it satisfies

(D5) *Lower range dominance*: $\mathscr{D}(X) \leq E[X] - \inf X$ for all X.

The importance of D5 will be elucidated in the context of the relationship between deviation measures and *coherent risk measures*.

Well-known examples of deviation measures include

(a) Deviation measures of $\mathscr{L}^p$-type: $\mathscr{D}(X) = \|X - E[X]\|_p$, $p \in [1, \infty]$, e.g., the *standard deviation* $\sigma(X) = \|X - E[X]\|_2$ and *mean absolute deviation* $\mathrm{MAD}(X) = \|X - E[X]\|_1$.
(b) Deviation measures of semi-$\mathscr{L}^p$ type: $\mathscr{D}_-(X) = \|[X - E[X]]_-\|_p$ and $\mathscr{D}_+(X) = \|[X - E[X]]_+\|_p$, $p \in [1, \infty]$, e.g., *standard lower semideviation* $\sigma_-(X) = \|[X - E[X]]_-\|_2$ and *standard upper semideviation* $\sigma_+(X) = \|[X - E[X]]_+\|_2$, and *lower* and *upper worst-case deviations*:

$$\mathscr{D}(X) = \|[X - E[X]]_-\|_\infty \equiv E[X] - \inf X,$$
$$\mathscr{D}'(X) = \|[X - E[X]]_+\|_\infty \equiv \sup X - E[X]$$

for a bounded random variable X.
(c) *CVaR deviation*: $\mathrm{CVaR}_\alpha^\Delta(X) = \mathrm{CVaR}_\alpha(X - E[X])$ for $\alpha \in [0, 1)$,[2] which can also be represented as the difference between the expected value of X and the average α-quantile of X:

---

[2] $\mathrm{CVaR}_1^\Delta(X) = -E[X] + E[X] = 0$ is not a deviation measure, since it vanishes for all r.v.'s (not only for constants).

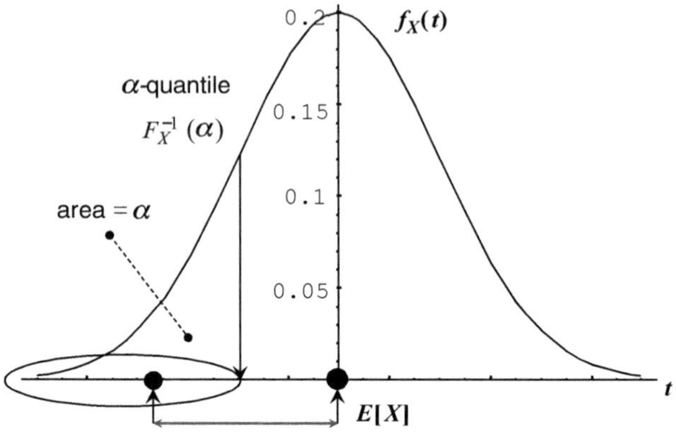

Fig. 2.1 CVaR deviation, $\mathrm{CVaR}_\alpha^\Delta(X)$, of a random variable X with PDF $f_X(t)$ is the difference between $E[X]$ and the average α-quantile

$$\mathrm{CVaR}_\alpha^\Delta(X) = E[X] - \frac{1}{\alpha}\int_0^\alpha q_X^+(s)\,ds, \tag{2.1.1}$$

see Fig. 2.1.

The measures $\mathscr{D}(X) = \|[X - E[X]]_-\|_p$, $p \in [1,\infty]$, and $\mathscr{D}(X) = \mathrm{CVaR}_\alpha^\Delta(X)$ are lower range dominated.[3]

Proposition 4 in [42] shows that deviation measures can be readily constructed out of given deviation measures $\mathscr{D}_1,\ldots,\mathscr{D}_m$ by the following two operations:

$$\mathscr{D}(X) = \sum_{k=1}^m \lambda_k\, \mathscr{D}_k(X), \qquad \sum_{k=1}^m \lambda_k = 1, \quad \lambda_k > 0, \quad k = 1,\ldots,m, \tag{2.1.2}$$

and

$$\mathscr{D}(X) = \max\{\mathscr{D}_1(X),\ldots,\mathscr{D}_m(X)\}. \tag{2.1.3}$$

In both cases, $\mathscr{D}(X)$ is lower range dominated if each $\mathscr{D}_k(X)$ is lower range dominated. For example, taking $\mathscr{D}_k(X) = \mathrm{CVaR}_{\alpha_k}^\Delta(X)$ with $\alpha_k \in (0,1)$ in (2.1.2) and (2.1.3), we obtain mixed-CVaR deviation

$$\mathscr{D}(X) = \sum_{k=1}^m \lambda_k\, \mathrm{CVaR}_{\alpha_k}^\Delta(X), \qquad \sum_{k=1}^m \lambda_k = 1, \quad \lambda_k > 0, \quad k = 1,\ldots,m, \tag{2.1.4}$$

and *max-CVaR deviation*

$$\mathscr{D}(X) = \max\{\mathrm{CVaR}_{\alpha_1}^\Delta(X),\ldots,\mathrm{CVaR}_{\alpha_m}^\Delta(X)\}.$$

---

[3]Indeed, $\|[X - E[X]]_-\|_p \le \|[X - E[X]]_-\|_\infty = E[X] - \inf X$ for $p \in [1,\infty]$, and $\mathrm{CVaR}_\alpha^\Delta(X) = E[X] + \mathrm{CVaR}_\alpha(X) \le E[X] - \inf X$.

In fact, mixed CVaR-deviation can be readily defined for continuously distributed λ as

$$\mathscr{D}(X) = \int_0^1 \mathrm{CVaR}_\alpha^\Delta(X)\, d\lambda(\alpha), \qquad \int_0^1 d\lambda(\alpha) = 1, \quad \lambda(\alpha) \geq 0, \tag{2.1.5}$$

which for $\lambda(\alpha) = \alpha^2$ is the product of $E[X]$ and the *Gini index* of X.[4]

Proposition 5 in [42] proves that if $\int_0^1 \alpha^{-1} d\lambda(\alpha) < \infty$, the mixed-CVaR deviation (2.1.5) can be represented in the equivalent form

$$\mathscr{D}(X) = \int_0^1 \mathrm{VaR}_\alpha(X - E[X])\phi(\alpha)\, d\alpha, \qquad \phi(\alpha) = \int_\alpha^1 \frac{d\lambda(s)}{s}, \tag{2.1.6}$$

where $\phi(\alpha)$ is a left-continuous and nonincreasing function with $\phi(0+) < \infty$, $\phi(1-) = 0$, and $\int_0^1 \phi(\alpha)d\alpha = 1$ and plays a role similar to that of a *dual utility* function in [47, 62].

A generalization of (2.1.5) is *worst-case mixed-CVaR deviation*

$$\mathscr{D}(X) = \sup_{\lambda \in \Xi} \int_0^1 \mathrm{CVaR}_\alpha^\Delta(X)\, d\lambda(\alpha) \tag{2.1.7}$$

for some set Ξ of weighting nonnegative measures λ on $(0, 1)$ with $\int_0^1 d\lambda(\alpha) = 1$.

The deviation measures (2.1.2)–(2.1.7) provide a powerful modeling tool for customizing agent's risk preferences, where the weights $\lambda_1, \ldots, \lambda_n$ and the weighting measure $\lambda(\alpha)$ can be considered as discrete and continuous *risk profiles*, respectively.

2.2 Risk Envelopes and Risk Identifiers

Deviation measures have dual characterization in terms of *risk envelopes* $\mathscr{Q} \subset \mathscr{L}^2(\Omega)$ defined by the properties

(Q1) $\mathscr{Q}$ is nonempty, closed, and convex.
(Q2) For every nonconstant X, there is some $Q \in \mathscr{Q}$ such that $E[XQ] < E[X]$.
(Q3) $E[Q] = 1$ for all $Q \in \mathscr{Q}$.

There is a one-to-one correspondence between deviation measures and risk envelopes [42, Theorem 1]:

[4]The Gini index of a random variable X is defined by $G_X = 1 - 2\int_0^1 L(\alpha)\, d\alpha$, where $L(\alpha) = \frac{1}{\mu_X}\int_0^\alpha q_X^+(s)\, ds$ is the *Lorenz curve* with $\mu_X = E[X]$, so that $\mu_X G_X = \int_0^1 \mathrm{CVaR}_\alpha^\Delta(X)\, d\alpha^2$.

$$\begin{aligned} \mathscr{D}(X) &= E[X] - \inf_{Q \in \mathscr{Q}} E[XQ], \\ \mathscr{Q} &= \{Q \in \mathscr{L}^2(\Omega) \mid \mathscr{D}(X) \geq E[X] - E[XQ] \text{ for all } X\}, \end{aligned} \tag{2.2.1}$$

and a deviation measure $\mathscr{D}$ is lower range dominated if and only if the corresponding risk envelope $\mathscr{Q}$ satisfies

(Q4) $Q \geq 0$ for all $Q \in \mathscr{Q}$.

Each Q of $\mathscr{Q}$ satisfying Q1–Q4 can be regarded as a "distortion" of the probability measure $\mathbb{P}$ that yields an "alternative" probability measures $\mathbb{P}_Q$ through $d\mathbb{P}_Q = Q\,d\mathbb{P}$. In this case, the corresponding deviation measure $\mathscr{D}(X) = E[X] - \inf_{Q \in \mathscr{Q}} E[XQ] \equiv E_{\mathbb{P}}[X] - \inf_{Q \in \mathscr{Q}} E_{\mathbb{P}_Q}[X]$ is the difference of what is expected under $\mathbb{P}$ and under the worst-case "alternative" probability distribution.

The elements of $\mathscr{Q}$ at which $E[XQ]$ attains infimum for a given X are called *risk identifiers* for X:

$$\mathscr{Q}(X) = \arg\min_{Q \in \mathscr{Q}} E[XQ].$$

They are those elements of $\mathscr{Q}$ that track the downside of X as closely as possible.

The second relationship in (2.2.1) implies that the set of risk identifiers for X with respect to a deviation measure $\mathscr{D}$ is determined by

$$\mathscr{Q}_{\mathscr{D}}(X) = \{Q \in \mathscr{Q} \mid \mathscr{D}(X) = E[(E[X] - X)Q] \equiv \mathrm{Cov}(-X, Q)\}.$$

From the optimization perspective, $\mathscr{Q}_{\mathscr{D}}(X)$ is closely related to *subdifferential* $\partial\mathscr{D}(X)$ of a deviation measure $\mathscr{D}$ at X, which is the set of *subgradients* $Z \in \mathscr{L}^2(\Omega)$ such that

$$\mathscr{D}(Y) \geq \mathscr{D}(X) + E[(Y - X)Z] \quad \text{for all} \quad Y \in \mathscr{L}^2(\Omega).$$

Namely,

$$\partial\mathscr{D}(X) = 1 - \mathscr{Q}_{\mathscr{D}}(X);$$

see [44, Proposition 1].

For σ, σ_-, MAD, $E[X] - \inf X$, and $\mathrm{CVaR}_\alpha^\Delta$, the corresponding risk envelopes $\mathscr{Q}$ and sets of risk identifiers $\mathscr{Q}_{\mathscr{D}}(X)$ are given by

(i) Standard deviation $\mathscr{D}(X) = \sigma(X)$:

$$\mathscr{Q} = \{Q \mid E[Q] = 1,\ \sigma(Q) \leq 1\},$$
$$\mathscr{Q}_\sigma(X) = \left\{1 - \frac{X - E[X]}{\sigma(X)}\right\}.$$

(ii) Standard lower semideviation $\mathscr{D}(X) = \sigma_-(X)$:

$$\mathcal{Q} = \{Q \mid E[Q] = 1,\ \|Q - \inf Q\|_2 \le 1\},$$
$$\mathcal{Q}_{\sigma_-}(X) = \left\{1 - \frac{E[Y] - Y}{\sigma_-(X)}\right\},$$

where $Y = [E[X] - X]_+$.

(iii) Mean absolute deviation $\mathcal{D}(X) = \mathrm{MAD}(X)$:

$$\mathcal{Q} = \{Q \mid E[Q] = 1,\ \sup Q - \inf Q \le 2\},$$
$$\mathcal{Q}_{\mathrm{MAD}}(X) = \{Q = 1 + E[Z] - Z \mid Z \in \mathrm{sign}\,[X - E[X]]\}.$$

(iv) Lower worst-case deviation $\mathcal{D}(X) = E[X] - \inf X$:

$$\mathcal{Q} = \{Q \mid E[Q] = 1,\ Q \ge 0\},$$
$$\mathcal{Q}_{\mathcal{D}}(X) = \{Q \mid E[Q] = 1,\ Q \ge 0,\ Q(\omega) = 0 \text{ when } X(\omega) > \inf X\}.$$

(v) CVaR deviation $\mathcal{D}(X) = \mathrm{CVaR}_\alpha^\Delta(X)$:

$$\mathcal{Q} = \{Q \mid E[Q] = 1,\ 0 \le Q \le 1/\alpha\} \tag{2.2.2}$$

and $\mathcal{Q}_{\mathrm{CVaR}_\alpha^\Delta}(X)$ is the set of elements Q such that $E[Q] = 1$ and

$$Q(\omega)\begin{cases} = \alpha^{-1} & \text{on } \{\omega \mid X(\omega) < -\mathrm{VaR}_\alpha(X)\},\\ \in [0, \alpha^{-1}] & \text{on } \{\omega \mid X(\omega) = -\mathrm{VaR}_\alpha(X)\},\\ = 0 & \text{on } \{\omega \mid X(\omega) > -\mathrm{VaR}_\alpha(X)\}.\end{cases} \tag{2.2.3}$$

Observe that for σ and σ_-, $\mathcal{Q}_{\mathcal{D}}(X)$ is a *singleton*, whereas (iv) is a particular case of (v) for $\alpha = 1$. See [44] for other examples of $\mathcal{Q}$ and $\mathcal{Q}_{\mathcal{D}}(X)$.

For the deviation measures (2.1.2) and (2.1.3), the risk envelopes are given by Proposition 4 in [42]:

$$\mathcal{Q} = \begin{cases} \text{closure of } \sum_{k=1}^{m} \lambda_k\, \mathcal{Q}_k \text{ for (2.1.2)},\\ \text{closure of the convex hull of } \cup_{k=1}^{m} \mathcal{Q}_k \text{ for (2.1.3)},\end{cases}$$

where $\mathcal{Q}_1, \ldots, \mathcal{Q}_m$ are the risk envelopes for the deviation measures $\mathcal{D}_1, \ldots, \mathcal{D}_m$. This result and the formula (2.2.2) imply that the risk envelope for the mixed-CVaR deviation (2.1.4) is determined by

$$\mathcal{Q} = \text{closure of} \sum_{k=1}^{m} \lambda_k\, Q_k, \quad \text{where} \quad E[Q_k] = 1,\ 0 \le Q_k \le \frac{1}{\alpha_k},\ k = 1, \ldots, m. \tag{2.2.4}$$

Risk identifiers along with risk envelopes are instrumental in formulating optimality conditions and devising optimization procedures in applications involving deviation measures. For example, if X is discretely distributed with $\mathbb{P}[X = x_k] =$

p_k, $k = 1, \ldots, n$, then with the risk envelope representation (2.2.2), the CVaR deviation (2.1.1) is readily restated as a linear program

$$\mathrm{CVaR}_\alpha^\Delta(X) = E[X] - \min_{q_1,\ldots,q_n} \left\{ \sum_{k=1}^{n} q_k p_k x_k \,\middle|\, q_k \in [0, \alpha^{-1}],\ \sum_{k=1}^{n} q_k p_k = 1 \right\},$$

whereas for the same X, mixed-CVaR deviation (2.1.4) with (2.2.4) can be represented by

$$\sum_{i=1}^{m} \lambda_i \, \mathrm{CVaR}_{\alpha_i}^\Delta(X) = E[X] - \min_{q_{ik}} \left\{ \sum_{i,k=1}^{m,n} \lambda_i \, q_{ik} p_k x_k \,\middle|\, q_{ik} \in [0, \alpha_i^{-1}],\ \sum_{k=1}^{n} q_{ik} p_k = 1 \right\}.$$

2.3 Averse Measures of Risk

Averse measures of risk [41, 42] are functionals $\mathcal{R} : \mathcal{L}^2(\Omega) \to (-\infty; \infty]$ satisfying[5]

(R1) *Risk aversion*: $\mathcal{R}(c) = -c$ for constants c, but $\mathcal{R}(X) > E[-X]$ for nonconstant X.
(R2) *Positive homogeneity*: $\mathcal{R}(\lambda X) = \lambda \mathcal{R}(X)$ when $\lambda > 0$.
(R3) *Subadditivity*: $\mathcal{R}(X + Y) \le \mathcal{R}(X) + \mathcal{R}(Y)$ for all X and Y.
(R4) *Lower semicontinuity*: set $\{X \in \mathcal{L}^2(\Omega) | \mathcal{R}(X) \le c\}$ is closed for all $c < \infty$.

Axiom R1 requires an additional explanation: $\mathcal{R}(c) = -c$ implies $\mathcal{R}(E[X]) = -E[X]$, so that the other condition $\mathcal{R}(X) > E[-X]$ can be restated as $\mathcal{R}(X) > \mathcal{R}(E[X])$ for $X \neq c$, which is the risk aversion property in terms of $\mathcal{R}$ (a risk-averse agent always prefers $E[X]$ over nonconstant X). Also, it follows from $\mathcal{R}(c) = -c$ and R3 that

$$\mathcal{R}(X + c) = R(X) - c,$$

which is called *translation invariance* in [2] (this property should not be confused with insensitivity to constant shift for deviation measures).

Averse measures of risk are closely related to *coherent risk measures* in the sense of Artzner et al. [2]. They share three main properties: subadditivity, positive homogeneity, and translation invariance. However, coherent risk measures $\mathcal{R}$ are monotone, i.e., satisfy

---

[5] In [41], these measures were originally called *strict expectation bounded risk measures*, and then in the subsequent work of [46], they were renamed averse measures of risk to reflect the concept more accurately.

(R5) *Monotonicity*: $\mathscr{R}(X_1) \geq \mathscr{R}(X_2)$ when $X_1 \leq X_2$

and may fail to satisfy R1 for nonconstant X. In other words, averse measures of risk are not required to be monotone, while coherent risk measures are not, in general, risk averse. Functionals $\mathscr{R} : \mathscr{L}^2(\Omega) \to (-\infty; \infty]$ that satisfy all R1–R5 are called *coherent averse measures of risk*; see [41, 42] for details.

Theorem 2 in [42] establishes a one-to-one correspondence between deviation measures and averse measures of risk through the relationships:

$$\mathscr{R}(X) = \mathscr{D}(X) - E[X], \qquad \mathscr{D}(X) = \mathscr{R}(X - E[X]) \tag{2.3.1}$$

and shows that $\mathscr{R}$ is a coherent averse measure of risk if and only if the corresponding deviation measure $\mathscr{D}$ is lower range dominated, i.e., satisfies D5. This result provides a simple recipe for constructing averse measures of risk:

(a) *Risk measures of $\mathscr{L}^p$-type*

$$\mathscr{R}(X) = \lambda \|X - E[X]\|_p - E[X], \qquad p \in [1, \infty], \quad \lambda > 0,$$

e.g., $\mathscr{R}(X) = \lambda\, \sigma(X) - E[X]$ and $\mathscr{R}(X) = \lambda\, \mathrm{MAD}(X) - E[X]$.

(b) *Risk measures of semi-$\mathscr{L}^p$ type*

$$\mathscr{R}(X) = \lambda \|[X - E[X]]_-\|_p - E[X], \qquad p \in [1, \infty], \quad \lambda > 0,$$

e.g., $\mathscr{R}(X) = \lambda\, \sigma_-(X) - E[X]$.

(c) *Risk measures of CVaR-type*: (i) $\mathscr{R}(X) = \mathrm{CVaR}_\alpha(X)$; (ii) *mixed CVaR*

$$\mathscr{R}(X) = \int_0^1 \mathrm{CVaR}_\alpha(X) d\lambda(\alpha),$$

where $\lambda(\alpha)$ is a measure such that $\lambda(\alpha) \geq 0$ and $\int_0^1 d\lambda(\alpha) = 1$; and (iii) *worst-case mixed CVaR*

$$\mathscr{R}(X) = \sup_{\lambda \in \Xi} \int_0^1 \mathrm{CVaR}_\alpha(X) d\lambda(\alpha),$$

where Ξ is a set of weighting nonnegative measures λ on $(0, 1)$ with $\int_0^1 d\lambda(\alpha) = 1$. These measures correspond to CVaR deviation (2.1.1), mixed CVaR-deviation (2.1.5), and worst-case mixed-CVaR deviation (2.1.7), respectively.

Among these, risk measures of CVaR-type and of semi-$\mathscr{L}^p(\Omega)$ type with $\lambda \in (0, 1]$ are coherent. Also, mixed CVaR can be equivalently represented in the form

$$\mathscr{R}(X) = \int_0^1 \mathrm{VaR}_\alpha(X)\, \phi(\alpha)\, d\alpha,$$

where ϕ is a left continuous and nonincreasing function with $\phi(0+) < \infty, \phi(1-) = 0$, and $\int_0^1 \phi(\alpha)d\alpha = 1$ and is also called *risk profile*; see [42, Proposition 5].

Another major implication of Theorem 2 in [42] is that all optimization procedures available for deviation measures can be readily applied to averse measures of risk. In particular, $\mathscr{R}$ and $\mathscr{D}$ corresponding through (2.3.1) have the same risk envelope and the same set of risk identifiers, so that

$$\begin{aligned} \mathscr{R}(X) &= -\inf_{Q\in\mathscr{Q}} E[XQ], \\ \mathscr{Q} &= \{Q \in \mathscr{L}^2(\Omega) \mid \mathscr{R}(X) \geq -E[XQ] \text{ for all } X\}, \end{aligned} \tag{2.3.2}$$

where in addition, $\mathscr{R}$ is coherent if and only if the corresponding risk envelope $\mathscr{Q}$ satisfies Q4.

Averse measures of risk can be also characterized in terms of *acceptance sets* $\mathscr{A} \subset \mathscr{L}^2(\Omega)$ defined by

(A1) $\mathscr{A}$ is closed and contains positive constants c.
(A2) $0 \in \mathscr{A}$, and $\lambda X \in \mathscr{A}$ whenever $X \in \mathscr{A}$ and $\lambda > 0$.
(A3) $X + Y \in \mathscr{A}$ for any $X \in \mathscr{A}$ and $Y \in \mathscr{A}$.
(A4) $E[X] > 0$ for every $X \not\equiv 0$ in $\mathscr{A}$.

A one-to-one correspondence between averse measures of risk $\mathscr{R}$ and acceptance sets $\mathscr{A}$ is established by:

$$\mathscr{A} = \{X \mid \mathscr{R}(X) \leq 0\}, \qquad \mathscr{R}(X) = \inf\{c \mid X + c \in \mathscr{A}\}; \tag{2.3.3}$$

see [41, Theorem 2]. In other words, *a random variable X is accepted or belongs to an acceptance set $\mathscr{A}$ if its risk is nonpositive, i.e., $\mathscr{R}(X) \leq 0$, whereas $\mathscr{R}(X)$ can be interpreted as the minimal cash reserve (possibly negative) making $X + \mathscr{R}(X)$ acceptable.* In addition, $\mathscr{R}$ is coherent if and only if $\mathscr{A}$ contains all nonnegative X. With the relationships (2.3.3), examples of acceptance sets for averse measures of risk are straightforward:

(a) $\mathscr{A} = \{X \mid \lambda \|X - E[X]\|_p \leq E[X]\}$ for the risk measures of $\mathscr{L}^p$-type with $p \in [1, \infty]$, $\lambda > 0$.
(b) $\mathscr{A} = \{X \mid \lambda \|[X - E[X]]_-\|_p \leq E[X]\}$ for the risk measures of semi-$\mathscr{L}^p$ type with $p \in [1, \infty]$, $\lambda > 0$.
(c) $\mathscr{A} = \{X \mid \mathrm{CVaR}_\alpha(X) \leq 0\}$ for $\mathscr{R}(X) = \mathrm{CVaR}_\alpha(X)$, $\alpha \in [0, 1]$.

In view of (2.3.3), $\mathscr{R}$ can be also interpreted as *$\mathscr{A}$-effective infimum* of X: $\mathscr{R}(X) = -\mathscr{A}\text{-}\inf X = \inf_{X+c\in\mathscr{A}} c$, and $\mathscr{D}$ corresponding to $\mathscr{R}$ through (2.3.1) can be restated as $\mathscr{D}(X) = E[X] - \mathscr{A}\text{-}\inf X$. This provides an interesting interpretation of $\mathscr{D}$: for each X, $\mathscr{D}(X)$ is the least upper bound of the difference between what is expected and what is accepted under given $\mathscr{A}$. For detailed discussion of these and other issues concerning averse measures of risk, the reader may refer to [41, 42].

2.4 Error Measures

Error measures [41, 42] are functionals $\mathscr{E} : \mathscr{L}^2(\Omega) \to [0, \infty]$ satisfying

(E1) *Nonnegativity*: $\mathscr{E}(0) = 0$, but $\mathscr{E}(X) > 0$ for $X \neq 0$; also, $\mathscr{E}(c) < \infty$ for constants c.
(E2) *Positive homogeneity*: $\mathscr{E}(\lambda X) = \lambda \mathscr{E}(X)$ when $\lambda > 0$.
(E3) *Subadditivity*: $\mathscr{E}(X + Y) \leq \mathscr{E}(X) + \mathscr{E}(Y)$ for all X and Y.
(E4) *Lower semicontinuity*: set $\{X \in \mathscr{L}^2(\Omega) | \mathscr{E}(X) \leq c\}$ is closed for all $c < \infty$.

For example, $\mathscr{L}^p$ norms $\|X\|_p$ with $p \geq 1$ are error measures. However, like deviation measures and averse measures of risk, error measures are not required to be symmetric $\mathscr{E}(-X) \neq \mathscr{E}(X)$ to allow treating gains and losses differently.

An example of *asymmetric* error measure is given by

$$\mathscr{E}_{a,b,p}(X) = \|a\, X_+ + b\, X_-\|_p, \qquad a > 0, \quad b > 0, \quad 1 \leq p \leq \infty. \tag{2.4.1}$$

Another one is the asymmetric mean absolute error (2.0.3), which is used in the quantile regression [26] in Sect. 6.4. Observe that for $a = 1$ and $b = 1$, (2.4.1) simplifies to $\|X\|_p$, whereas for $p = 1$, $a = 1$, and $b = 1/\alpha - 1$, it reduces to (2.0.3).

An error measure $\mathscr{E}$ is *nondegenerate* if there exists $\delta > 0$ such that $\mathscr{E}(X) \geq \delta\, |E[X]|$ for all X. For example, (2.4.1) and (2.0.3) are both nondegenerate error measures with $\delta = \min\{a, b\}$ and $\delta = \min\{1, 1/\alpha - 1\}$, respectively; see [46].

Functionals $\mathscr{D}$, $\mathscr{R}$, and $\mathscr{E}$ share the same three properties: positive homogeneity, subadditivity, and lower semicontinuity. The only difference comes from axioms D1, R1, and E1 on how the functionals treat constants. In fact, any two of D1, R1, and E1 are incompatible, i.e., there is no functional satisfying any two of these three axioms.

Unlike the relationships (2.3.1), there is no one-to-one correspondence between deviation measures and error measures. Nonetheless, each error measure $\mathscr{E}$ yields a deviation measure through *penalties relative to expectation*

$$\mathscr{D}(X) = \mathscr{E}(X - E[X]), \tag{2.4.2}$$

and if $\mathscr{E}$ is nondegenerate, it furnishes another deviation through *error projection*

$$\mathscr{D}(X) = \inf_{c \in \mathbb{R}} \mathscr{E}(X - c), \tag{2.4.3}$$

which is called *the deviation of X projected from $\mathscr{E}$*; see Theorem 2.1 in [46]. A solution to (2.4.3) is *the statistic of X associated with $\mathscr{E}$*

$$\mathscr{S}(X) = \arg\min_{c \in \mathbb{R}} \mathscr{E}(X - c), \tag{2.4.4}$$

which, in general, is an interval $[\mathcal{S}^-(X), \mathcal{S}^+(X)]$ of constants with $\mathcal{S}^-(X) = \min\{c \,|c \in \mathcal{S}(X)\}$ and $\mathcal{S}^+(X) = \max\{c \,|c \in \mathcal{S}(X)\}$ and has the following properties:

$$\mathcal{S}(X - c) = \mathcal{S}(X) - c \quad \text{for any constant } c,$$
$$\mathcal{S}(\lambda\, X) = \lambda\, \mathcal{S}(X) \quad \text{for any constant } \lambda > 0.$$

Well-known examples of the relationships (2.4.3) and (2.4.4) include

$\mathcal{E}(X)$	$\mathcal{D}(X)$	$\mathcal{S}(X)$
$\|X\|_2$	$\sigma(X)$	$E[X]$
$\|X\|_1$	$\|X - \text{med}\,(X)\|_1$	$\text{med}\,(X)$
$\mathcal{E}_\alpha(X) = \frac{1}{\alpha} E[\alpha\, X_+ + (1-\alpha)\, X_-]$	$\text{CVaR}_\alpha^\Delta(X)$	$q_X(\alpha) = [q_X^-(\alpha), q_X^+(\alpha)]$

where med (X) is the median of X (possibly an interval). Observe that for $\mathcal{E}(X) = \|X\|_2$, deviations (2.4.2) and (2.4.3) coincide, whereas for $\mathcal{E}(X) = \|X\|_1$, they are different.

If $\mathcal{E}_1, \ldots, \mathcal{E}_l$ are nondegenerate error measures that project to deviations $\mathcal{D}_1, \ldots, \mathcal{D}_l$, respectively, then, for any weights $\lambda_1 > 0, \ldots, \lambda_l > 0$ with $\sum_{k=1}^{l} \lambda_k = 1$,

$$\mathcal{E}(X) = \inf_{\substack{c_1,\ldots,c_l \\ \lambda_1 c_1 + \cdots + \lambda_l c_l = 0}} \sum_{k=1}^{l} \lambda_k\, \mathcal{E}_k(X - c_k)$$

is a nondegenerate error measure, which projects to the deviation measure

$$\mathcal{D}(X) = \sum_{k=1}^{l} \lambda_k \mathcal{D}_k(X)$$

with the associated statistic

$$\mathcal{S}(X) = \sum_{k=1}^{l} \lambda_k \mathcal{S}_k(X);$$

see [46, Theorem 2.2]. As an immediate consequence of this result, we restate Example 2.6 from [46].

Example 2.1 (mixed quantiles and mixed-CVaR deviation). For any choice of probability thresholds $\alpha_k \in (0, 1)$ and weights $\lambda_1 > 0, \ldots, \lambda_l > 0$ with $\sum_{k=1}^{l} \lambda_k = 1$,

$$\mathscr{E}(X) = E[X] + \inf_{\substack{c_1,\dots,c_l \\ \lambda_1 c_1+\dots+\lambda_l c_l=0}} \sum_{k=1}^{l} \frac{\lambda_k}{\alpha_k} E[\max\{0, c_k - X\}] \tag{2.4.5}$$

is a nondegenerate error measure (called *mixed quantile error measure*), which projects to the mixed-CVaR deviation (2.1.4) with the associated statistic

$$\mathscr{S}(X) = \sum_{k=1}^{l} \lambda_k\, q_X(\alpha_k), \qquad q_X(\alpha_k) = [q_X^-(\alpha_k), q_X^+(\alpha_k)]. \tag{2.4.6}$$

For a given deviation measure $\mathscr{D}$, a nondegenerate error measure can be obtained by *inverse projection*

$$\mathscr{E}(X) = \mathscr{D}(X) + |E[X]|,$$

which through (2.4.3) projects back to $\mathscr{D}$ with the associated statistic $\mathscr{S}(X) = E[X]$ [46, Example 2.5]. For example, $\mathscr{E}(X) = \sigma_-(X) + |E[X]|$ is an error measure that projects to $\sigma_-(X)$. There could be more than one error measure projecting to the same deviation measure, e.g., $\mathscr{E}(X) = \|X\|_2$ and $\mathscr{E}(X) = \|X - E[X]\|_2 + |E[X]|$ both project to $\mathscr{D}(X) = \sigma(X)$, and an arbitrary nondegenerate error measure $\mathscr{E}$ can be modified as

$$\mathscr{E}'(X) = \inf_{c\in\mathbb{R}} \mathscr{E}(X - c) + |E[X]| = \mathscr{E}(X - E[X]) + |E[X]|$$

to have $E[X]$ as the associated statistic (in this case, $\mathscr{E}'(X) \equiv \mathscr{E}'(X - E[X]) + |E[X]|$).

For a given error measure $\mathscr{E}(X)$, the representations (2.4.2) and (2.4.3) along with the relationships (2.3.1) provide two ways for constructing (different) averse measures of risk

$$\mathscr{R}(X) = \mathscr{E}(X - E[X]) - E[X]$$

and

$$\mathscr{R}(X) = \inf_{c\in\mathbb{R}} \mathscr{E}(X - c) - E[X].$$

Remarkably, for the asymmetric mean absolute error (2.0.3), the last formula can be restated as $\mathscr{R}(X) = \inf_{c\in\mathbb{R}} E[\alpha^{-1}[c - X]_+ - c]$, which is (1.4.4) with negative sign that defines CVaR. For other examples of relationships between $\mathscr{D}$, $\mathscr{R}$, and $\mathscr{E}$, in particular for the error measure corresponding to the mixed-CVaR deviation (2.1.4), see [41, 42, 46].

Chapter 3
Probabilistic Inequalities

In various statistical decision problems dealing with safety and reliability, risk is often interpreted as the probability of a dread event or disaster, and minimizing the probability of a highly undesirable event is known as the *safety-first principle* [50]. If the CDF of X is either unknown or complex, the probability in question can be estimated through more simple characteristics such as mean and standard deviation of X, for example, by Markov's and Chebyshev's inequalities. Also, if the probability depends on decision variables, then, in general, an optimization problem, in which it is either minimized or constrained, is nonconvex. In this case, the probability can be estimated by an appropriate probabilistic inequality, and then the optimization problem can be approximated by a convex one; see, e.g., [3, 32].

3.1 Basic Probabilistic Inequalities

Markov's inequality is one of the basic and yet most important probabilistic inequalities. It is given by

$$\mathbb{P}[|X| \geq a] \leq \frac{E[|X|]}{a}, \qquad a > 0. \tag{3.1.1}$$

Its proof is straightforward:

$$|X| \geq |X| \, I_{\{|X| \geq a\}} \geq a \, I_{\{|X| \geq a\}},$$

where $I_{\{|X| \geq a\}}$ is the indicator function equal to 1 if the condition in the curly brackets is true and equal to 0 otherwise. Consequently,

$$E[|X|] \geq E[a \, I_{\{|X| \geq a\}}] \equiv a \, \mathbb{P}[|X| \geq a].$$

M. Zabarankin and S. Uryasev, *Statistical Decision Problems: Selected Concepts and Portfolio Safeguard Case Studies*, Springer Optimization and Its Applications 85,
DOI 10.1007/978-1-4614-8471-4_3,

Despite its simplicity, (3.1.1) is the major source for obtaining other well-known basic probabilistic inequalities:

- Replacing $|X|$ and a in (3.1.1) by $|X|^p$ and a^p, respectively, where $p > 0$, we obtain

$$\mathbb{P}[|X| \geq a] \leq \left(\frac{\|X\|_p}{a}\right)^p, \qquad a > 0.$$

- For $Y = \ln|X|$ and $b = \ln a$, (3.1.1) is transformed into the inequality for estimating the probability of the right tail $Y \geq b$:

$$\mathbb{P}[Y \geq b] \leq E\left[\mathrm{e}^{Y-b}\right].$$

- Substituting $(X - E[X])^2$ for $|X|$ and $a\,\sigma^2(X)$ for a in (3.1.1), we obtain *Chebyshev's inequality*:

$$\mathbb{P}[|X - \mu| \geq a] \leq \frac{\sigma^2(X)}{a^2}, \qquad a > 0,$$

which evaluates the probability of how significantly a random variable X deviates from its expected value $\mu = E[X]$ in terms of the standard deviation $\sigma(X)$.

There is also *one-sided Chebyshev's inequality*, also called *Cantelli's inequality*, that estimates the probability of X either not to exceed a given threshold τ:

$$\mathbb{P}[X \leq \tau] \leq \frac{\sigma^2(X)}{\sigma^2(X) + (\mu - \tau)^2}, \qquad \tau \leq \mu, \tag{3.1.2}$$

or not to drop below the threshold τ:

$$\mathbb{P}[X \geq \tau] \leq \frac{\sigma^2(X)}{\sigma^2(X) + (\tau - \mu)^2}, \qquad \tau \geq \mu. \tag{3.1.3}$$

Though (3.1.2) and (3.1.3) do not follow from (3.1.1) as simply as two-sided Chebyshev's inequality, their proof still relies on Markov's inequality. For example, for (3.1.2), consider

$$\mathbb{P}[X \leq \tau] = \mathbb{P}[t - X \geq t - \tau] \leq \mathbb{P}\left[(t - X)^2 \geq (t - \tau)^2\right] \leq \frac{E\left[(t - X)^2\right]}{(t - \tau)^2},$$

where t is an arbitrary real number greater than τ and where the last inequality follows from (3.1.1). Since the inequality $\mathbb{P}[X \leq \tau] \leq E\left[(t - X)^2\right]/(t - \tau)^2$ holds for any $t \geq \tau$, setting the derivative of its right-hand side with respect to t to zero, we obtain that $t^* = \mu + \sigma^2(X)/(\mu - \tau) > \tau$ is the minimizer, and $\mathbb{P}[X \leq \tau] \leq E\left[(t^* - X)^2\right]/(t^* - \tau)^2$ reduces to (3.1.2). Observe that the condition $t^* \geq \tau$ holds if and only if $\mu \geq \tau$.

As an immediate application of (3.1.2), consider a probabilistic constraint (also known as *chance constraint*)

$$\mathbb{P}[X \le \tau] \le \alpha, \tag{3.1.4}$$

where $\alpha \in (0, 1)$ is given. Then one-sided Chebyshev's inequality (3.1.2) implies that (3.1.4) is guaranteed to hold if

$$E[X] - \sigma(X)\sqrt{\alpha^{-1} - 1} \ge \tau, \tag{3.1.5}$$

which is a simple and frequently used condition provided that $E[X]$ and $\sigma(X)$ are either known or easy to estimate.

Also, since $\mathbb{P}[X \le \tau] \equiv F_X(\tau)$, by integrating (3.1.2) with respect to τ, we obtain an estimate for $F_X^{(2)}(\tau) = \int_{-\infty}^{\tau} F_X(s)\,ds$:

$$F_X^{(2)}(\tau) \le \sigma(X)\left(\frac{\pi}{2} - \arctan\left(\frac{\mu - \tau}{\sigma(X)}\right)\right), \qquad \tau \le \mu.$$

Markov's and Chebyshev's inequalities are sources for other many remarkable probabilistic relationships and are proved to be invaluable in decision problems with insufficient statistical data; see, e.g., [3, 32].

Another useful and frequently used probabilistic inequality is that of Kolmogorov. Suppose $X_1, \ldots, X_n$ are a sequence of independent random variables such that $E[X_k] = 0$ and $\sigma(X_k) < \infty$, $k = 1, \ldots, n$, and let $S_k = \sum_{j=1}^{k} X_j$. Then *Kolmogorov's inequality* estimates the probability of $\max\limits_{1\le k\le n} |S_k|$ to exceed a threshold a in terms of $\sigma(S_n)$:

$$\mathbb{P}\left[\max_{1\le k\le n} |S_k| \ge a\right] \le \frac{1}{a}\sum_{k=1}^{n} \sigma^2(X_k) \equiv \frac{\sigma^2(S_n)}{a}. \tag{3.1.6}$$

Chebyshev's and Kolmogorov's inequalities can be improved if the standard deviation is replaced by another deviation measure. Next sections present generalizations of Chebyshev's and Kolmogorov's inequalities and discuss application of generalized inequalities in statistical decision problems.

3.2 Chebyshev's Inequalities with Deviation Measures

3.2.1 One-Sided Chebyshev's Inequalities

The problem of generalizing one-sided Chebyshev's inequality for *law-invariant* deviation measures,[1] e.g., σ, σ_-, MAD, and $\mathrm{CVaR}_\alpha^\Delta$, is formulated as follows: for

[1]A deviation measure $\mathcal{D}(X)$ is law invariant if for any two random variables X_1 and X_2 having the same probability distribution, $\mathcal{D}(X_1) = \mathcal{D}(X_2)$, i.e., if $\mathcal{D}(X)$ depends only on the probability distribution of X.

law-invariant $\mathscr{D} : \mathscr{L}^p(\Omega) \to [0, \infty]$, $1 \le p < \infty$, and fixed $a > 0$, find a function $g_{\mathscr{D}}(d)$ such that

$$\mathbb{P}[X \le \mu - a] \le g_{\mathscr{D}}(\mathscr{D}(X)) \quad \text{for all } X \in \mathscr{L}^p(\Omega), \tag{3.2.1}$$

where $\mu = E[X]$, under the conditions: (i) $g_{\mathscr{D}}$ is independent of the distribution of X and (ii) $g_{\mathscr{D}}$ is the least upper bound in (3.2.1), i.e., for every $d > 0$, there is a random variable X such that (3.2.1) becomes the equality with $\mathscr{D}(X) = d$.

The inequality (3.2.1) can be reformulated as an optimization problem

$$\begin{aligned} u_{\mathscr{D}}(\delta) = & \inf_{X \in \mathscr{L}^p(\Omega)} \mathscr{D}(X) \\ & \text{subject to} \quad X \in \mathscr{U} = \{X \mid E[X] = 0,\ \mathbb{P}[X \le -a] \ge \delta\}, \end{aligned} \tag{3.2.2}$$

with the function $g_{\mathscr{D}}$ determined by

$$g_{\mathscr{D}}(d) = \sup_{\delta \in (0,1)} \{\delta \mid u_{\mathscr{D}}(\delta) \le d\}; \tag{3.2.3}$$

see [17] for details. Proposition 3 in [17] proves that (3.2.2) is equivalent to minimizing $\mathscr{D}$ over a subset of $\mathscr{U}$, whose elements are undominated random variables with respect to *convex ordering*,[2] and that a solution of (3.2.2) is the random variable $X^*(\delta)$ assuming only two values $-a$ and $\delta a/(1 - \delta)$ with the probabilities δ and $1 - \delta$, respectively, i.e.,

$$\mathbb{P}[X^*(\delta) = -a] = \delta, \qquad \mathbb{P}\left[X^*(\delta) = \frac{\delta a}{1-\delta}\right] = 1 - \delta.$$

Thus,

$$u_{\mathscr{D}}(\delta) = \mathscr{D}(X^*(\delta)),$$

and if $u_{\mathscr{D}}$ has the inverse $u_{\mathscr{D}}^{-1}$, then (3.2.3) implies that $g_{\mathscr{D}}(d) = u_{\mathscr{D}}^{-1}(d)$.

For σ, MAD, σ_-, and $\text{CVaR}_\alpha^\Delta$, the function $u_{\mathscr{D}}(\delta)$ and its inverse, which is $g_{\mathscr{D}}(d)$, are given by

$$\begin{aligned} u_\sigma(\delta) &= \sigma(X^*(\delta)) = a\sqrt{\frac{\delta}{1-\delta}}, & g_\sigma(d) &= u_\sigma^{-1}(d) = \frac{d^2}{a^2 + d^2}, \\ u_{\text{MAD}}(\delta) &= \text{MAD}(X^*(\delta)) = 2a\delta, & g_{\text{MAD}}(d) &= u_{\text{MAD}}^{-1}(d) = \frac{d}{2a}, \\ u_{\sigma_-}(\delta) &= \sigma_-(X^*(\delta)) = a\sqrt{\delta}, & g_{\sigma_-}(d) &= u_{\sigma_-}^{-1}(d) = \frac{d^2}{a^2}, \end{aligned}$$

[2] X dominates Y with respect to convex ordering if $E[f(X)] \ge E[f(Y)]$ for any convex function $f : \mathbb{R} \mapsto \mathbb{R}$, which is equivalent to the conditions $E[X] = E[Y]$ and $\int_{-\infty}^{x} F_X(t)dt \ge \int_{-\infty}^{x} F_Y(t)dt$ for all $x \in \mathbb{R}$, where F_X and F_Y are CDFs of X and Y, respectively.

and

$$u_{\mathrm{CVaR}_\alpha^\Delta}(\delta) = \mathrm{CVaR}_\alpha^\Delta(X^*(\delta)) = \begin{cases} a & \text{if } \alpha < \delta, \\ \frac{a\delta(1-\alpha)}{\alpha(1-\delta)} & \text{if } \alpha \ge \delta, \end{cases}$$

$$g_{\mathrm{CVaR}_\alpha^\Delta}(d) = u^{-1}_{\mathrm{CVaR}_\alpha^\Delta}(d) = \frac{\alpha\, d}{\alpha\, d + (1-\alpha)\, a} \quad \text{if } d < a.$$

Let $a = \mu - \tau > 0$. Then for $\mathscr{D} = \sigma$, one-sided Chebyshev's inequality (3.2.1) with $g_\sigma(d)$ reduces to (3.1.2), whereas for MAD, σ_-, and $\mathrm{CVaR}_\alpha^\Delta$, (3.2.1) with the corresponding $g_{\mathscr{D}}(d)$ takes the form

$$\mathbb{P}[X \le \tau] \le \frac{\mathrm{MAD}(X)}{2(\mu - \tau)}, \qquad \tau < \mu, \tag{3.2.4}$$

$$\mathbb{P}[X \le \tau] \le \frac{\sigma_-(X)^2}{(\mu - \tau)^2}, \qquad \tau < \mu, \tag{3.2.5}$$

and

$$\mathbb{P}[X \le \tau] \le \frac{\alpha\, \mathrm{CVaR}_\alpha^\Delta(X)}{\alpha\, \mathrm{CVaR}_\alpha^\Delta(X) + (1-\alpha)(\mu - \tau)}, \qquad \tau < -\mathrm{CVaR}_\alpha(X). \tag{3.2.6}$$

The inequality (3.2.6) can be restated in terms of the average α-quantile $\overline{q}_X(\alpha)$:

$$\mathbb{P}[X \le \tau] \le \frac{\alpha\,(\mu - \overline{q}_X(\alpha))}{\mu - \tau - \alpha\,(\overline{q}_X(\alpha) - \tau)}, \qquad \tau < \overline{q}_X(\alpha). \tag{3.2.7}$$

In (3.2.6) and (3.2.7), α is a parameter. What α provides the best estimate? This question can be readily answered by recasting the right-hand side of (3.2.7) in the form

$$1 - \frac{(1-\alpha)(\mu - \tau)}{\mu - (1-\alpha)\tau - q_X^{(2)}(\alpha)}$$

and by observing that its second term (fraction) is a decreasing function of α. Consequently, given that $\tau < \mu$, the right-hand sides in (3.2.6)–(3.2.7) attain minimum at the lowest possible α, which, however, does not exist since $\overline{q}_X(\alpha) > \tau$. To circumvent this technicality, the lowest α can be defined as

$$\alpha_X^*(\tau) = \inf\{\alpha \,|\, \overline{q}_X(\alpha) > \tau\} = \sup\{\alpha \,|\, \overline{q}_X(\alpha) \le \tau\},$$

where the second equality holds due to $\overline{q}_X(\alpha)$ being a nondecreasing function of α. Since for given X and τ, the right-hand side in (3.2.7) is a continuous function of α and approaches $\alpha_X^*(\tau)$ as $\alpha \to \alpha_X^*(\tau)+$, it can be shown that Chebyshev's inequality (3.2.7) holds in the limit case of $\alpha \to \alpha_X^*(\tau)+$:

$$\mathbb{P}[X \leq \tau] \leq \alpha_X^*(\tau). \tag{3.2.8}$$

In some reliability applications, $\alpha_X^*(\tau)$ is called *buffered probability*; see [38].

Now we are ready to address the question of approximating the probabilistic constraint (3.1.4) by one-sided Chebyshev's inequality with MAD, σ_-, and $\mathrm{CVaR}_\alpha^\Delta$. Namely, the constraint (3.1.4) is guaranteed to hold if the right-hand side of any of the inequalities (3.2.4), (3.2.5), or (3.2.8) is less than or equal to α, which is equivalent to the conditions

$$E[X] - \frac{1}{2\alpha}\mathrm{MAD}(X) \geq \tau, \tag{3.2.9}$$

$$E[X] - \frac{1}{\sqrt{\alpha}}\sigma_-(X) \geq \tau, \tag{3.2.10}$$

$$\alpha_X^*(\tau) \leq \alpha \quad \text{with} \quad \overline{q}_X(\alpha_X^*(\tau)) \equiv \tau,$$

respectively, where the last one reduces to

$$\overline{q}_X(\alpha) \geq \tau. \tag{3.2.11}$$

In fact, the constraint (3.1.4) is equivalent to

$$q_X^+(\alpha) \geq \tau, \tag{3.2.12}$$

so that the left-hand sides of (3.1.5), (3.2.9), (3.2.10), and (3.2.11) are upper bounds of $q_X^+(\alpha)$. Recall that $q_X^+(\alpha)$ is a nonconvex function of X, and optimization problems involving the constraint (3.2.12) are difficult to solve, whereas all these four bounds are concave functions of X. So, which bound among these four is tighter?

As an illustration, the four bounds are evaluated for a standard normal random variable Z:

$$E[Z] - \sigma(Z)\sqrt{\alpha^{-1} - 1} = -\sqrt{\alpha^{-1} - 1},$$

$$E[Z] - \frac{1}{2\alpha}\mathrm{MAD}(Z) = -\frac{1}{\sqrt{2\pi}\,\alpha},$$

$$E[Z] - \frac{1}{\sqrt{\alpha}}\sigma_-(Z) = -\frac{1}{\sqrt{2\alpha}},$$

$$\overline{q}_Z(\alpha) = -\frac{1}{\sqrt{2\pi}\,\alpha}\exp\left(-\frac{1}{2}\left(\Phi^{-1}(\alpha)\right)^2\right),$$

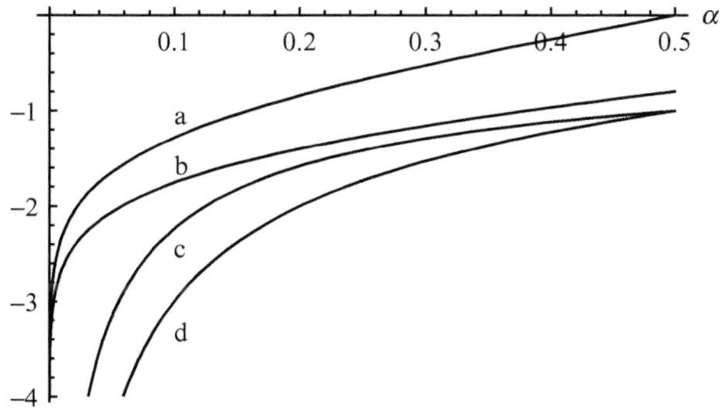

Fig. 3.1 Comparison of $q_Z^+(\alpha)$ (curve a) with $\overline{q}_Z(\alpha)$ (curve b), $-\sqrt{\alpha^{-1}-1}$ (curve c), and $-1/\sqrt{2\alpha}$ (curve d) for a standard normal random variable Z for $\alpha \in (0, 1/2]$

where $\Phi^{-1}(\alpha)$ is the inverse of the CDF of Z: $\Phi(t) = \frac{1}{\sqrt{2\pi}}\int_{-\infty}^{t} e^{-s^2/2}ds$. Figure 3.1 shows $q_Z^+(\alpha) = \Phi^{-1}(\alpha)$, $-\sqrt{\alpha^{-1}-1}$, $-1/\sqrt{2\alpha}$, and $\overline{q}_Z(\alpha)$ for $\alpha \in (0, 1/2]$. For very small α, $\overline{q}_Z(\alpha)$ is close to $q_Z^+(\alpha)$, whereas $-\sqrt{\alpha^{-1}-1}$, which corresponds to σ, is a quite conservative bound over the whole range.

The following general result holds. Let

$$\mathscr{C}_\sigma = \left\{X \;\middle|\; E[X] - \sigma(X)\sqrt{\alpha^{-1}-1} \geq \tau\right\}, \qquad \mathscr{C}_{\overline{q}_X} = \{X \mid \overline{q}_X(\alpha) \geq \tau\}$$

be the feasible sets of X for the constraints (3.1.5) and (3.2.11), respectively. Then

$$\mathscr{C}_\sigma \subseteq \mathscr{C}_{\overline{q}_X};$$

see the proof in [32]. It shows that the constraint (3.2.11) yields a larger feasible set than (3.1.5) does. Also, for a discretely distributed random variable X, $\overline{q}_X(\alpha)$ in (3.2.11) can be reformulated as the linear program (1.4.5), which is attractive from the computational perspective; see [32]. However, the advantage of (3.1.5) is in its simplicity and possibility to obtain closed-form analytical solutions; see [32].

3.2.2 *Two-Sided Chebyshev's Inequalities*

The problem for generalizing two-sided Chebyshev's inequality for an arbitrary law-invariant deviation measure is formulated similarly to (3.2.1); see [17] for details.

As in (3.2.1), let $\mu = E[X]$ and $a > 0$. Two-sided Chebyshev's inequality with MAD, σ_-, and $\mathrm{CVaR}_\alpha^\Delta$ is given by

$$\mathbb{P}[|X-\mu| \geq a] \leq \frac{\mathrm{MAD}(X)}{a}, \tag{3.2.13}$$

$$\mathbb{P}[|X-\mu| \geq a] \leq \begin{cases} \frac{16}{9}d + \frac{1}{9}, & d \geq \frac{1}{20}, \\ \frac{\sqrt{d^2+4d}-d}{2}, & d < \frac{1}{20}, \end{cases} \tag{3.2.14}$$

where $d = \sigma_-^2(X)/a^2$, and

$$\mathbb{P}[|X-\mu| \geq a] \leq \begin{cases} \frac{d}{1+d}, & d \leq \frac{1/2-\alpha}{1/2+\alpha}, \\ \frac{1}{2} + \alpha d - \sqrt{\alpha(1-d)(1-\alpha(1+d))}, & d \in \left[\frac{1/2-\alpha}{1/2+\alpha}, 1\right), \\ 1, & d \geq 1, \end{cases} \tag{3.2.15}$$

where $d = \mathrm{CVaR}_\alpha^\Delta(X)/a$ and $\alpha \leq 1/2$. For $\alpha \geq 1/2$, two-sided Chebyshev's inequality with $\mathrm{CVaR}_\alpha^\Delta$ is similar to (3.2.15) and can be obtained by using the relation $\alpha \cdot \mathrm{CVaR}_\alpha^\Delta(X) = (1-\alpha) \cdot \mathrm{CVaR}_{1-\alpha}^\Delta(-X)$.

Let $\mathscr{D}(X)$ be *max-CVaR deviation* defined by

$$\mathscr{D}(X) = \max\left\{\mathrm{CVaR}_\alpha^\Delta(X), \mathrm{CVaR}_\alpha^\Delta(-X)\right\}, \tag{3.2.16}$$

then for $\alpha \leq 1/2$, two-sided Chebyshev's inequality with (3.2.16) takes the form

$$\mathbb{P}[|X-\mu| \geq a] \leq \begin{cases} \frac{2\alpha\mathscr{D}(X)}{a}, & \mathscr{D}(X) < a, \\ 1, & \mathscr{D}(X) \geq a. \end{cases} \tag{3.2.17}$$

3.3 Kolmogorov's Inequalities with Deviation Measures

Proposition 12 in [17] generalizes Kolmogorov's inequality (3.1.6) with arbitrary low-invariant deviation measure in place of standard deviation. In fact, for same deviation measure, Kolmogorov's and Chebyshev's inequalities are closely related.

Let $X_1, \ldots, X_n$ be independent random variables such that $E[X_k] = 0$ and $\mathscr{D}(X_k) < \infty$, $k = 1, \ldots, n$, and let $S_k = \sum_{j=1}^k X_j$. Then *Kolmogorov's inequality* with MAD, σ_-, $\mathrm{CVaR}_\alpha^\Delta$, and max CVaR deviation (3.2.16) is given by

$$\mathbb{P}\left[\max_{1\le k\le n} |S_k| \ge a\right] \le \begin{cases} \text{the right-hand side of (3.2.13) with } X = S_n \text{ for MAD}, \\ \text{the right-hand side of (3.2.14) with } X = S_n \text{ for } \sigma_-, \\ \text{the right-hand side of (3.2.15) with } X = S_n \text{ for CVaR}_\alpha^\Delta, \\ \text{the right-hand side of (3.2.17) with } X = S_n \text{ for (3.2.16)}. \end{cases} \tag{3.3.1}$$

Classical Kolmogorov's inequality (3.1.6) fits the pattern in (3.3.1) as well.

Example 10 in [17] shows that Kolmogorov's inequality (3.3.1) with max-CVaR deviation (3.2.16) provides the best bound in the following probability estimation problem. Let S_t with $t = 0, 1, 2, \ldots$ and $S_0 = 0$ be a discrete-time martingale such that the increments $X_t = S_t - S_{t-1}$ are independent and identically distributed random variables with mean 0 and finite variance σ^2. S_t can model the logarithm of the exchange rate of two currencies or the logarithm of the rate of return of a stock.

Suppose S_t is the logarithm of the exchange rate of two currencies, and we are interested in estimating $\mathbb{P}[|S_t| < a \text{ for all } t \le n]$. For illustration, let $a = \sigma\sqrt{n}$. For sufficiently large n, the distribution of S_n is approximately normal, i.e., $S_n \sim N\left(0, n\sigma^2\right)$. Then classical Kolmogorov's inequality (3.1.6) implies

$$\mathbb{P}\left[\max_{1\le t\le n} |S_t| \ge a\right] \le \frac{\sigma^2(S_n)}{a^2} = 1,$$

which is uninformative, whereas Kolmogorov's inequality (3.3.1) with MAD yields

$$\mathbb{P}\left[\max_{1\le t\le n} |S_t| \ge a\right] \le \frac{\text{MAD}(S_n)}{a} = \sqrt{\frac{2}{\pi}} \approx 0.8,$$

where $\text{MAD}(S_n)/a = \text{MAD}(S_n/a)$ follows from the positive homogeneity property of deviation measures and $\text{MAD}(S_n/a) = \sqrt{2/\pi}$ follows from the fact that $S_n/a \sim N(0,1)$. However, Kolmogorov's inequality (3.3.1) with max-CVaR deviation (3.2.16) furnishes even better bound:

$$\mathbb{P}\left[\max_{1\le t\le n} |S_t| \ge a\right] \le 2\alpha_0 \approx 0.764,$$

where α_0 is such that $\text{CVaR}_{\alpha_0}^\Delta(Z) = 1$ for $Z \sim N(0,1)$. Remarkably, the last bound cannot be improved further by using another deviation measure in Kolmogorov's inequality; see Example 10 in [17] for details.

Part II
Statistical Decision Problems

Chapter 4
Maximum Likelihood Method

A classical problem in the statistical decision theory is to estimate the probability distribution of a random vector X given its independent observations $x_1, \ldots, x_n$. Often it is assumed that the probability distribution comes from some family of functions parametrized by a set of parameters $\theta_1, \ldots, \theta_m$, so that in this case, the problem is reduced to estimating $\theta_1, \ldots, \theta_m$ and is called *parametric estimation*. However, if no specific family of distributions is assumed, i.e., the probability distribution can not be completely defined by a finite number of parameters, the problem is called *nonparametric estimation*.

In both parametric and nonparametric estimations, there are several approaches to determine the probability distribution in question: the *maximum likelihood principle*, *maximum entropy principle*, and the *minimum relative entropy principle* (or the *principle of minimum discrimination information*). These principles are closely related and are the subject of this chapter and the next one.

4.1 Maximum Likelihood Principle

The *maximum likelihood principle* is arguably the most widely used principle in the statistical decision theory. It maximizes the probability of observing $x_1, \ldots, x_n$ either with respect to $\theta_1, \ldots, \theta_m$ in parametric estimation or with respect to the probability distribution of X in nonparametric estimation. The latter is a calculus of variations problem.

4.1.1 Parametric Estimation

We begin with considering the maximum likelihood principle in parametric estimation. First, suppose X is discretely distributed with a probability mass function

M. Zabarankin and S. Uryasev, *Statistical Decision Problems: Selected Concepts and Portfolio Safeguard Case Studies*, Springer Optimization and Its Applications 85,
DOI 10.1007/978-1-4614-8471-4_4,

$P_X(t;\theta_1,\ldots,\theta_m)$. Then the probability of observing independent $x_1,\ldots,x_n$ is given by the product $\prod_{k=1}^n P_X(x_k;\theta_1,\ldots,\theta_m)$, and the maximum likelihood principle leads to the optimization problem

$$\max_{(\theta_1,\ldots,\theta_m)\in\mathbb{R}^m} \prod_{k=1}^{n} P_X(x_k;\theta_1,\ldots,\theta_m). \tag{4.1.1}$$

The problem (4.1.1) is often reformulated as

$$\max_{(\theta_1,\ldots,\theta_m)\in\mathbb{R}^m} \sum_{k=1}^{n} \ln P_X(x_k;\theta_1,\ldots,\theta_m), \tag{4.1.2}$$

which has certain technical advantages compared to (4.1.1) and corresponds to the so-called maximum log-likelihood principle. If $P_X(t;\theta_1,\ldots,\theta_m)$ is continuously differentiable with respect to $\theta_1,\ldots,\theta_m$, the first-order necessary optimality conditions for (4.1.2) are given by

$$\sum_{k=1}^{n} \frac{1}{P_X(x_k;\theta_1,\ldots,\theta_m)} \frac{\partial P_X(x_k;\theta_1,\ldots,\theta_m)}{\partial\theta_j} = 0, \qquad j = 1,\ldots,m, \tag{4.1.3}$$

which, in general, is a system of m nonlinear equations for determining $\theta_1,\ldots,\theta_m$. Optimal $\theta_1,\ldots,\theta_m$ will be denoted by $\hat{\theta}_1,\ldots,\hat{\theta}_m$ and are collectively called *maximum likelihood estimator*.

Example 4.1 (Poisson distribution). If $X : \Omega \to \mathbb{N}\cup\{0\}$ is assumed to have the Poisson distribution $\mathbb{P}[X = k] = \lambda^k e^{-\lambda}/k!$, $k = 0,1,2,3,\ldots$, with unknown λ, then $m = 1$ and $\theta_1 = \lambda$, and (4.1.3) yields $\hat{\lambda} = \frac{1}{n}\sum_{k=1}^n x_k$, which is the sample mean.

If X is continuously distributed with a PDF $f_X(t;\theta_1,\ldots,\theta_m)$, in which $\theta_1,\ldots,\theta_m$ are unknown parameters to be estimated based on independent observations $x_1,\ldots,x_n$, the approach remains the same. The only difference is that in (4.1.1)–(4.1.3), $P_X(t;\theta_1,\ldots,\theta_m)$ is replaced by $f_X(t;\theta_1,\ldots,\theta_m)$.

Example 4.2 (normal distribution: mean estimation). Suppose $X : \Omega \to \mathbb{R}$ is normally distributed with unknown mean μ and given standard deviation σ, so that $m = 1$ and $\theta_1 = \mu$. Then (4.1.3) with $f_X(x_k,\mu) = \frac{1}{\sqrt{2\pi}\sigma}e^{-\frac{(x_k-\mu)^2}{2\sigma^2}}$ in place of $P_X(x_k;\theta_1)$ implies $\hat{\mu} = \frac{1}{n}\sum_{k=1}^n x_k$.

In both Examples 4.1 and 4.2, the optimal parameters λ and μ are given by the sample mean. This is not accidental. Indeed, in both cases, λ and μ represent theoretical expectation of X, and the *law of large numbers* (in strong form) states that the sample average $\frac{1}{n}\sum_{k=1}^n x_k$ of independent $x_1,\ldots,x_n$ converges to the expected value of X almost surely.

Example 4.3 (normal distribution: variance estimation). Suppose $X : \Omega \to \mathbb{R}$ is again normally distributed but, this time, with given mean μ and unknown variance σ^2, i.e., $\theta_1 = \sigma^2$. Then (4.1.3) with $f_X(x_k, \sigma^2) = \frac{1}{\sqrt{2\pi}\sigma} e^{-\frac{(x_k-\mu)^2}{2\sigma^2}}$ in place of $P_X(x_k; \theta_1)$ yields $\hat{\sigma}^2 = \frac{1}{n}\sum_{k=1}^{n}(x_k - \mu)^2$, which coincides with the sample variance.

The next example illustrates estimation of two unknown parameters.

Example 4.4 (normal distribution: mean and variance estimation). Suppose $X : \Omega \to \mathbb{R}$ is normally distributed with unknown mean μ and unknown variance σ^2, i.e., $m = 2$ and $\theta_1 = \mu$, $\theta_2 = \sigma^2$. Then (4.1.3) with $f_X(x_k, \mu, \sigma^2) = \frac{1}{\sqrt{2\pi}\sigma} e^{-\frac{(x_k-\mu)^2}{2\sigma^2}}$ in place of $P_X(x_k; \theta_1, \theta_2)$ implies

$$\hat{\mu} = \frac{1}{n}\sum_{k=1}^{n} x_k \quad \text{and} \quad \hat{\sigma}^2 = \frac{1}{n}\sum_{k=1}^{n}(x_k - \hat{\mu})^2,$$

which are the sample mean and sample variance, respectively.[1]

Observe that if $f_X(t; \theta_1, \ldots, \theta_m)$ is not continuously differentiable with respect to $\theta_1, \ldots, \theta_m$, then the maximum likelihood principle should be used in the form (4.1.2) or even in (4.1.1) if, in addition, the support of $f_X(t; \theta_1, \ldots, \theta_m)$ is finite. This is demonstrated by the next example.

Example 4.5 (uniform distribution). Suppose X is uniformly distributed on $[0, a]$ with $a > 0$ being unknown, and suppose all $x_1, \ldots, x_n$ are nonnegative. In this case, $f_X(t, a) = 1/a$ for $t \in [0, a]$ and $f_X(t, a) = 0$ for $t \notin [0, a]$, so that $\ln f_X(t, a)$ does not exist when $t \notin [0, a]$ and $\ln f_X(t, a)$ is not differentiable at $t = 0$ and $t = a$. The maximum likelihood principle in the form (4.1.1) implies that the likelihood function does not vanish only if $a \geq \max\{x_1, \ldots, x_n\}$. On the other hand, the likelihood function decreases with an increase of a. Consequently, optimal a should assume its lower bound which is $\max\{x_1, \ldots, x_n\}$. In other words,

$$\hat{a} = \max_{1 \leq k \leq n} x_k$$

is the maximum likelihood estimator, which is simply the sample's upper bound.

In fact, maximum likelihood estimators are *consistent*, which means that as $n \to \infty$, they converge to the true values of the estimated parameters, for example, to the mean in Examples 4.1–4.4 and to the variance in Examples 4.3–4.4.

[1] In this case, sample variance is also defined as $\frac{1}{n-1}\sum_{k=1}^{n}(x_k - \hat{\mu})^2$, which is *unbiased estimator*.

4.1.2 Nonparametric Estimation

Now let $X : \Omega \to \mathbb{R}$ be a discretely distributed random variable, but this time no specific distribution family for X is assumed. Let $y_1, \ldots, y_l$ be distinct observations with y_k appearing n_k times in $x_1, \ldots, x_n$, so that $n_1 + \ldots + n_l = n$. In this case, the distribution of X can be represented by unknown probabilities $p_k, k = 1, \ldots, l$, such that $\mathbb{P}[X = y_k] = p_k > 0$ and $\sum_{k=1}^{l} p_k = 1$. The problem of estimating $p_1 \ldots, p_l$ through the maximum likelihood principle is similar to (4.1.2):

$$\max_{p_1,\ldots,p_l} \sum_{k=1}^{l} \frac{n_k}{n} \ln p_k$$

$$\text{subject to} \sum_{k=1}^{l} p_k = 1, \quad p_k > 0, \quad k = 1, \ldots, l, \tag{4.1.4}$$

where the multiplier $1/n$ in the objective function is introduced for convenience.

In contrast to (4.1.2), (4.1.4) is a constrained optimization problem, which can be solved by the Lagrange multipliers technique. The Lagrangian of (4.1.4) is given by

$$L(p_1, \ldots, p_l, \lambda) = \sum_{k=1}^{n} \frac{n_k}{n} \ln p_k - \lambda \left(\sum_{k=1}^{l} p_k - 1 \right),$$

where $\lambda \in \mathbb{R}$ is the Lagrange multiplier corresponding to the constraint $\sum_{k=1}^{l} p_k = 1$. Setting the derivatives of $L(p_1, \ldots, p_l, \lambda)$ with respect to $p_1, \ldots, p_l$ and λ to zero, we obtain optimal p_k:

$$p_k = \frac{n_k}{n}, \qquad k = 1, \ldots, l, \tag{4.1.5}$$

which coincide with the empirical distribution of the sample $x_1, \ldots, x_n$.

Remark 4.1. Observe that the constraints $p_1 > 0, \ldots, p_l > 0$ are satisfied automatically, since when $p_k \to 0+$ for some k, the objective function decreases without bound. In fact, we can define $\ln p = -\infty$ at $p = 0$ and replace the constraints $p_1 > 0, \ldots, p_l > 0$ in (4.1.4) by $p_1 \geq 0, \ldots, p_l \geq 0$. Then the problem (4.1.4) is equivalent to minimizing the convex function $-\sum_{k=1}^{l} \frac{n_k}{n} \ln p_k$, which may take on $+\infty$, subject to the linear constraints $\sum_{k=1}^{l} p_k = 1$, $p_1 \geq 0, \ldots, p_l \geq 0$, so that the solution (4.1.5), found from the necessary optimality conditions, is the global maximum of (4.1.4) (see, e.g., Theorem 3.34 in [51]). This remark concerns similar problems in the rest of this chapter.

The problem (4.1.4) can be considered under an additional assumption that some moments of X are either given or known to be within certain bounds.

For a particular application, this additional information can be provided, for example, by experts or analysts. Specifically, we assume that j^{th} moment of X belongs to the interval $[a_j, b_j]$ with given a_j and b_j for $j = 1, \ldots, m$, i.e., $a_j \le E[X^j] \le b_j$, $j = 1, \ldots, m$; see [61]. In this case, the problem (4.1.4) is augmented with the additional constraints $a_j \le \sum_{k=1}^{l} p_k y_k^j \le b_j$, $j = 1, \ldots, m$:

$$\begin{aligned} \max_{p_1,\ldots,p_l} \quad & \sum_{k=1}^{l} \frac{n_k}{n} \ln p_k \\ \text{subject to} \quad & \sum_{k=1}^{l} p_k = 1, \quad p_k > 0, \quad k = 1, \ldots, l, \\ & a_j \le \sum_{k=1}^{l} p_k y_k^j \le b_j, \quad j = 1, \ldots, m. \end{aligned} \tag{4.1.6}$$

In general, (4.1.6) has no closed-form solution, but can be efficiently solved by numerical optimization techniques. The analysis of (4.1.6) simplifies considerably when $a_1 = b_1, \ldots, a_m = b_m$.

Example 4.6 (known mean). Suppose the mean of X is known to be μ. In this case, $m = 1$ and $a_1 = b_1 = \mu$, and the problem (4.1.6) takes the form

$$\begin{aligned} \max_{p_1,\ldots,p_l} \quad & \sum_{k=1}^{l} \frac{n_k}{n} \ln p_k \\ \text{subject to} \quad & \sum_{k=1}^{l} p_k = 1, \quad \sum_{k=1}^{l} p_k y_k = \mu, \quad p_k > 0, \quad k = 1, \ldots, l. \end{aligned} \tag{4.1.7}$$

The Lagrange multipliers technique yields

$$p_k = \frac{n_k}{n(1 + \lambda\,(y_k - \mu))}, \qquad k = 1, \ldots, l,$$

provided that the Lagrange multiplier $\lambda \in \mathbb{R}$, associated with the constraint $\sum_{k=1}^{l} p_k y_k = \mu$, satisfies

$$\begin{cases} \displaystyle\sum_{k=1}^{l} \frac{n_k}{n(1 + \lambda\,(y_k - \mu))} = 1, \\ \displaystyle\sum_{k=1}^{l} \frac{n_k y_k}{n(1 + \lambda\,(y_k - \mu))} = \mu, \\ 1 + \lambda\,(y_k - \mu) > 0, \qquad k = 1, \ldots, l. \end{cases}$$

If $l > 1$ and such λ exists, then λ is unique. Indeed, suppose λ_1 and λ_2 satisfy the above system and $\lambda_1 \neq \lambda_2$. Then the system implies that

$$\sum_{k=1}^{l} \frac{n_k(y_k - \mu)}{1 + \lambda_1 (y_k - \mu)} = 0, \qquad \sum_{k=1}^{l} \frac{n_k(y_k - \mu)}{1 + \lambda_2 (y_k - \mu)} = 0.$$

Subtracting the second equation from the first one, we obtain

$$(\lambda_2 - \lambda_1) \sum_{k=1}^{l} \frac{n_k(y_k - \mu)^2}{(1 + \lambda_1 (y_k - \mu))(1 + \lambda_2 (y_k - \mu))} = 0.$$

However, since $1 + \lambda_1 (y_k - \mu) > 0$ and $1 + \lambda_2 (y_k - \mu) > 0$ for all k (because $p_k > 0$) and $l > 1$, i.e., not all $x_1, \ldots, x_n$ coincide with μ, the above equation can hold only if $\lambda_1 = \lambda_2$, which contradicts the assumption that $\lambda_1 \neq \lambda_2$.

4.1.3 Estimation of Probability Distributions of Random Vectors

Similarly, the maximum likelihood principle can be used for estimating unknown probability distribution of a random vector $X : \Omega \to \mathbb{R}^l$ based on independent observations $x_1, \ldots, x_n$ and other available information. The only difference is that in (4.1.1)–(4.1.3), x_k is now a vector.

First we proceed with an example of parametric estimation.

Example 4.7 (multivariate normal distribution: mean estimation). Let $X : \Omega \to \mathbb{R}^l$ be a random vector with multivariate normally distributed components, unknown mean $\mu \in \mathbb{R}^l$, and given covariance matrix Λ. In this case, $x_k \in \mathbb{R}^l$, $k = 1, \ldots, n$, and $\theta = \mu$. Then (4.1.3) with

$$f_X(x_k, \mu) = \frac{1}{(2\pi)^{l/2}\sqrt{\det \Lambda}} \exp\left(-\frac{1}{2}(x_k - \mu)^\top \Lambda^{-1}(x_k - \mu)\right)$$

in place of $P_X(x_k; \theta_1, \ldots, \theta_l)$ implies

$$\hat{\mu} = \frac{1}{n} \sum_{k=1}^{n} x_k,$$

which generalizes the estimator obtained in Example 4.2.

As an illustration of nonparametric estimation, consider two random variables $X_1 : \Omega \to \mathbb{R}$ and $X_2 : \Omega \to \mathbb{R}$ with unknown joint discrete distribution.

Suppose $x_1 \in \mathbb{R}^2, \ldots, x_n \in \mathbb{R}^2$ are independent observations of $X = (X_1, X_2)$, where $x_k = (x_{k1}, x_{k2})$ with $x_{k1} \in \mathbb{R}$ and $x_{k2} \in \mathbb{R}$ being observations of X_1 and X_2, respectively. Also, suppose $y_1 \in \mathbb{R}^2, \ldots, y_l \in \mathbb{R}^2$ are distinct observations of X with y_k appearing n_k times in $x_1, \ldots, x_n$, so that $\sum_{k=1}^{l} n_k = n$. If the marginal distributions for X_1 and X_2 are known, then the problem of estimating the probability distribution of X with the maximum likelihood principle is formulated as follows.

Let $u_1, \ldots, u_{l_1}$ and $v_1, \ldots, v_{l_2}$ be distinct observations of X_1 and X_2, respectively, with n_{ij} being the number of times the pair (u_i, v_j) is observed in $x_1, \ldots, x_n$, where $\sum_{i=1}^{l_1} \sum_{j=1}^{l_2} n_{ij} = n$. Let $p_{ij} = \mathbb{P}[X_1 = u_i, X_2 = v_j] > 0$, $i = 1, \ldots, l_1$, $j = 1, \ldots, l_2$, be unknown joint distribution of X_1 and X_2 such that $\sum_{i=1}^{l_1} \sum_{j=1}^{l_2} p_{ij} = 1$, $\sum_{j=1}^{l_2} p_{ij} = \alpha_i$, $i = 1, \ldots, l_1$, and $\sum_{i=1}^{l_1} p_{ij} = \beta_j$, $j = 1, \ldots, l_2$, where $\alpha_i = \mathbb{P}[X_1 = u_i]$, $i = 1, \ldots, l_1$, and $\beta_j = \mathbb{P}[X_2 = v_j]$, $j = 1, \ldots, l_2$, are given marginals of X_1 and X_2, respectively. Observe that either $\sum_{j=1}^{l_2} p_{ij} = \alpha_i$, $i = 1, \ldots, l_1$, or $\sum_{i=1}^{l_1} p_{ij} = \beta_j$, $j = 1, \ldots, l_1$, imply $\sum_{i=1}^{l_1} \sum_{j=1}^{l_2} p_{ij} = 1$. Thus, the problem of likelihood maximization takes the form

$$
\begin{aligned}
\max_{p_{ij}} \quad & \sum_{i=1}^{l_1} \sum_{j=1}^{l_2} \frac{n_{ij}}{n} \ln p_{ij} \\
\text{subject to} \quad & \sum_{j=1}^{l_2} p_{ij} = \alpha_i, \quad i = 1, \ldots, l_1, \\
& \sum_{i=1}^{l_1} p_{ij} = \beta_j, \quad j = 1, \ldots, l_2, \\
& p_{ij} > 0, \quad i = 1, \ldots, l_1, \quad j = 1, \ldots, l_2,
\end{aligned}
\tag{4.1.8}
$$

where the multiplier $1/n$ in the objective function is introduced for convenience.

With the Lagrange multipliers $\lambda_1 \in \mathbb{R}, \ldots, \lambda_{l_1} \in \mathbb{R}$ and $\eta_1 \in \mathbb{R}, \ldots, \eta_{l_2} \in \mathbb{R}$, corresponding to the constraints on the marginals, the first-order necessary optimality conditions for (4.1.8) yield

$$
p_{ij} = \frac{n_{ij}}{n(\lambda_i + \eta_j)}, \quad i = 1, \ldots, l_1, \quad j = 1, \ldots, l_2,
$$

where $\lambda_1, \ldots, \lambda_{l_1}$ and $\eta_1, \ldots, \eta_{l_2}$ are to be found from the system

$$\begin{cases} \sum_{j=1}^{l_2} \dfrac{n_{ij}}{\lambda_i + \eta_j} = n\,\alpha_i, \quad i = 1, \ldots, l_1, \\ \sum_{i=1}^{l_1} \dfrac{n_{ij}}{\lambda_i + \eta_j} = n\,\beta_j, \quad j = 1, \ldots, l_2, \\ \lambda_i + \eta_j > 0, \quad i = 1, \ldots, l_1, \quad j = 1, \ldots, l_2. \end{cases}$$

In some cases, the problem (4.1.8) has closed-form solutions.

Example 4.8 (constraints on the marginal distribution of one random variable). If only the marginal distribution $\mathbb{P}[X_1 = u_i] = \alpha_i, i = 1, \ldots, l_1$, is given, then (4.1.8) has no constraints $\sum_{i=1}^{l_1} p_{ij} = \beta_j$, $j = 1, \ldots, l_2$, and the Lagrange multipliers $\lambda_1, \ldots, \lambda_{l_1}$ corresponding to $\sum_{j=1}^{l_2} p_{ij} = \alpha_i$, $i = 1, \ldots, l_1$, and probabilities p_{ij} are determined by

$$\lambda_i = \frac{1}{n\,\alpha_i} \sum_{j=1}^{l_2} n_{ij}, \qquad j = 1, \ldots, l_2,$$

$$p_{ij} = \frac{\alpha_i\, n_{ij}}{\sum_{j=1}^{l_2} n_{ij}}, \qquad i = 1, \ldots, l_1, \quad j = 1, \ldots, l_2.$$

If no information about marginals is available, the constraints on the marginals in (4.1.8) should be replaced by the constraint $\sum_{i=1}^{l_1} \sum_{j=1}^{l_2} p_{ij} = 1$, and in this case, likelihood maximization yields $p_{ij} = n_{ij}/n$, $i = 1, \ldots, l_1$, $j = 1, \ldots, l_2$, which is similar to (4.1.5).

Chapter 5
Entropy Maximization

The previous chapter showed that given independent observations of a random variable X, the probability distribution of X can be estimated based on the maximum likelihood principle. However, if no observations of X are available, but some integral characteristics of the distribution of X are known, for example, mean μ and standard deviation σ, the main principle for finding the distribution in question is, arguably, the one of *maximum entropy*. This principle, also known as *MaxEnt*, originated from the information theory and statistical mechanics (see [22]) and determines the "most unbiased" probability distribution for X subject to any constraints on X (prior information). Nowadays, it is widely used in financial engineering and statistical decision problems [4, 11, 56]. Estimation of probability distributions through entropy maximization and through relative entropy minimization subject to various constraints on unknown distributions is the subject of this chapter.

5.1 Shannon Entropy Maximization

A classical application of the maximum entropy principle in statistics is estimating the probability distribution of a random variable $X \in \mathscr{L}^m(\Omega)$ provided that the first m moments of X are known to be $\mu_1 \in \mathbb{R}, \ldots, \mu_m \in \mathbb{R}$:

$$\max_{X \in \mathscr{L}^m(\Omega)} S(X) \qquad \text{subject to} \quad E[X^k] = \mu_k, \quad k = 1, \ldots, m, \tag{5.1.1}$$

where $S(X)$ is the Shannon entropy of X.

If X is restricted to assume only n distinct values $x_1 \in \mathbb{R}, \ldots, x_n \in \mathbb{R}$ with nonnegative probabilities $p_1, \ldots, p_n$ summing to 1, then the problem (5.1.1) takes the form

M. Zabarankin and S. Uryasev, *Statistical Decision Problems: Selected Concepts and Portfolio Safeguard Case Studies*, Springer Optimization and Its Applications 85,
DOI 10.1007/978-1-4614-8471-4_5,

$$\max_{p_1,\dots,p_n} \quad -\sum_{j=1}^{n} p_j \ln p_j$$

$$\text{subject to} \quad \sum_{j=1}^{n} x_j^k \, p_j = \mu_k, \quad k = 0, 1, \dots, m,$$

$$p_j \ge 0, \quad j = 1, \dots, n, \tag{5.1.2}$$

whereas if X is continuously distributed with a PDF $f_X(t)$ on $[a,b]$ ($f_X(t) \equiv 0$ for $t \notin [a,b]$), it is formulated by

$$\max_{f_X(t)} \quad -\int_a^b f_X(t) \ln f_X(t)\, dt$$

$$\text{subject to} \quad \int_a^b t^k \, f_X(t)\, dt = \mu_k, \quad k = 0, 1, \dots, m,$$

$$f_X(t) \ge 0, \quad a \le t \le b. \tag{5.1.3}$$

In both (5.1.2) and (5.1.3), $\mu_0 = 1$. In view of Remark 1.1, some of $p_1, \dots, p_n$ in (5.1.2) can be zero, and $f_X(t)$ in (5.1.3) may vanish for some $t \in [a,b]$.

The problem (5.1.2) is convex with linear constraints, and its objective function is differentiable for any $p_1 > 0, \dots, p_n > 0$. If the equality (moment) constraints in (5.1.2) hold for some $\tilde{p}_1 > 0, \dots, \tilde{p}_n > 0$ (*Slater's constraint qualification condition*), then the optimality conditions for (5.1.2) yield

$$p_j = \exp\left(\sum_{k=0}^{m} \rho_k \, x_j^k\right), \qquad j = 1, \dots, n, \tag{5.1.4}$$

provided that there exist $\rho_0, \rho_1, \dots, \rho_m$ such that (5.1.4) satisfies the constraints in (5.1.2). Moreover, if exists, the solution (5.1.4) is the global maximum of (5.1.2); see, e.g., Theorem 3.34 in [51].

Similarly, the first-order necessary optimality conditions for (5.1.3) yield

$$f_X(t) = \exp\left(\sum_{k=0}^{m} \lambda_k \, t^k\right) I_{\{t \in [a,b]\}}, \tag{5.1.5}$$

provided that there exist $\lambda_0, \lambda_1, \dots, \lambda_m$ such that (5.1.5) satisfies the constraints in (5.1.3), where $I_{\{\cdot\}}$ is the indicator function equal to 1 if the condition in the curly brackets is true and equal to 0 otherwise. Boltzmann's theorem [10, Theorem 12.1.1] implies that if exists, the function (5.1.5) is the global maximum of (5.1.3).

We proceed with stating well-known examples of the maximum entropy distributions that follow from (5.1.4) to (5.1.5).

Example 5.1 (No prior information, $m = 0$). If no moments of a random variable X are known, then $m = 0$ and the maximum entropy distributions that solve (5.1.2) and (5.1.3) are uniform:

$$p_1 = \ldots = p_n = \frac{1}{n} \quad \text{in discrete case,}$$

$$f_X(t) = \frac{1}{b-a} I_{\{t \in [a,b]\}} \quad \text{in continuous case.} \tag{5.1.6}$$

In other words, the maximum entropy principle implies that without any information about a random variable X (either discretely or continuously distributed), all outcomes of X should be equally probable.

Example 5.2 (Known mean, $m = 1$). If it is only known that the mean of a random variable X is μ, then $m = 1$ and the maximum entropy distributions (5.1.4) and (5.1.5) take the form

$$p_k = \frac{e^{\rho x_k}}{\sum_{k=1}^{n} e^{\rho x_k}}, \quad k = 1, \ldots, n, \tag{5.1.7}$$

and

$$f_X(t) = \frac{\lambda\, e^{\lambda t}}{e^{\lambda b} - e^{\lambda a}} I_{\{t \in [a,b]\}}, \tag{5.1.8}$$

respectively, where ρ satisfies $\sum_{k=1}^{n}(x_k - \mu)e^{\rho x_k} = 0$ and $\lambda > 0$ is found from the equation $(\lambda(\mu - a) + 1)e^{\lambda a} = (\lambda(\mu - b) + 1)e^{\lambda b}$.

Corollary 5.1 (Known mean and semi-infinite support). *Example 5.2 implies that*

(a) If X is known to assume only integers starting from 1, i.e., $x_k = k$ with $k = 1, 2, \ldots$ and $\mu > 1$, then $e^{\rho} = 1 - 1/\mu$ and the maximum entropy distribution (5.1.7) *reduces to*

$$p_k = \frac{(\mu - 1)^{k-1}}{\mu^k}, \quad k = 1, 2, \ldots.$$

(b) If X is known to be continuously distributed on $[0, \infty)$ with $\mu > 0$, then $\lambda = -1/\mu$ and the maximum entropy solution (5.1.8) *simplifies to the exponential distribution $f_X(t) = e^{-t/\mu}/\mu$, $t \in [0, \infty)$.*

Example 5.3 (Known mean and standard deviation, $m = 2$). Given that a random variable X has mean μ and variance σ^2, $m = 2$ and the maximum entropy distribution (5.1.4) reduces to

$$p_k = \frac{\exp\left(\rho_1 x_k + \rho_2 x_k^2\right)}{\sum_{k=1}^n \exp\left(\rho_1 x_k + \rho_2 x_k^2\right)}, \quad k = 1, \ldots, n, \tag{5.1.9}$$

where ρ_1 and ρ_2 are found from the system

$$\begin{cases} \sum_{k=1}^{n} (x_k - \mu) \exp\left(\rho_1 x_k + \rho_2 x_k^2\right) = 0, \\ \sum_{k=1}^{n} (x_k^2 - \mu^2 - \sigma^2) \exp\left(\rho_1 x_k + \rho_2 x_k^2\right) = 0, \end{cases}$$

whereas (5.1.5) simplifies to

$$f_X(t) = \frac{\exp\left(\lambda_1 t + \lambda_2 t^2\right)}{\int_a^b \exp(\lambda_1 t + \lambda_2 t^2)\, dt} I_{\{t \in [a,b]\}} \tag{5.1.10}$$

with λ_1 and λ_2 found from the system

$$\begin{cases} \int_a^b (t - \mu) \exp\left(\lambda_1 t + \lambda_2 t^2\right) dt = 0, \\ \int_a^b \left(t^2 - \mu^2 - \sigma^2\right) \exp\left(\lambda_1 t + \lambda_2 t^2\right) dt = 0. \end{cases}$$

Unlike Example 5.2, Example 5.3 does not offer simplifications for discretely distributed random variables assuming infinitely many integer values either on $\mathbb{R}^+$ or $\mathbb{R}$. However, it yields an important corollary for a random variable continuously distributed on $\mathbb{R}$.

Corollary 5.2 (Continuously distributed random variable on $\mathbb{R}$ with given mean and variance). *If only mean μ and variance σ^2 of a continuously distributed random variable X on $\mathbb{R}$ are known, the maximum entropy PDF is given by $\frac{1}{\sqrt{2\pi}\sigma} \exp\left(-(t-\mu)^2/(2\sigma^2)\right)$. In other words, X is normally distributed with mean μ and variance σ^2.*

A generalization of the maximum entropy problem (5.1.3) is given by

$$\begin{aligned} \max_{f_X(t)} \quad & -\int_V f_X(t) \ln f_X(t)\, dt \\ \text{subject to} \quad & \int_V h_k(t)\, f_X(t)\, dt = a_k, \quad k = 1, \ldots, m, \\ & \int_V f_X(t)\, dt = 1, \qquad f_X(t) \ge 0, \quad t \in V, \end{aligned} \tag{5.1.11}$$

where V is a given closed support set $V \subseteq \mathbb{R}$ of $f_X(t)$, so that $f_X(t) \equiv 0$ for $t \notin V$; $h_1, \ldots, h_m$ are given measurable functions; and $a_1, \ldots, a_m$ are given constants.

Boltzmann's theorem [10, Theorem 12.1.1] shows that if there exist $\lambda_1, \ldots, \lambda_n$, and $c > 0$ such that the PDF

$$f_X(t) = c \exp\left(\sum_{k=1}^{m} \lambda_k h_k(t)\right) I_{\{t \in V\}} \tag{5.1.12}$$

satisfies the constraints in (5.1.11), then (5.1.12) is the global maximum of (5.1.11).

With arbitrary constraints on a random variable X, the maximum entropy problem has no closed-form solution regardless of whether X is distributed continuously or discretely. In this case, maximum entropy probability distributions are found by means of numerical optimization.

The next two examples present entropy maximization problems arising in *collateralized debt obligation* (CDO) pricing models [20].

Example 5.4 (Entropy maximization with no-arbitrage constraints). Suppose there are m CDO tranches and there are n scenarios for the *hazard rate* (λ in a Poisson process modeling default) with unknown probabilities $p_1, \ldots, p_n$. Let a_{ij} be the expected net payoff of tranche j in hazard rate scenario i. Under the no-arbitrage assumption, the expected net payoff of each CDO tranche over all hazard rate scenarios should be zero: $\sum_{i=1}^{n} a_{ij} p_i = 0$, $j = 1, \ldots, m$. If instead of a_{ij} we use the expected net payoffs $\underline{a}_{ij}$ and $\overline{a}_{ij}$ corresponding to ask and bid quotes for tranche j spread, then *no-arbitrage constraints* are given by $\sum_{i=1}^{n} \underline{a}_{ij} p_i \leq 0$ and $\sum_{i=1}^{n} \overline{a}_{ij} p_i \geq 0$, $j = 1, \ldots, m$, and the problem is to maximize the Shannon entropy with respect to the hazard rate scenario probabilities $p_1, \ldots, p_n$ subject to the no-arbitrage constraints:

$$\begin{aligned}
\max_{p_1,\ldots,p_n} \quad & -\sum_{i=1}^{n} p_i \ln p_i \\
\text{subject to} \quad & \sum_{i=1}^{n} p_i = 1, \quad p_i \geq 0, \quad i = 1, \ldots, n, \\
& \left.\begin{aligned} &\sum_{i=1}^{n} \underline{a}_{ij} p_i \leq 0, \quad j = 1, \ldots, m, \\ &\sum_{i=1}^{n} \overline{a}_{ij} p_i \geq 0, \quad j = 1, \ldots, m. \end{aligned}\right\} \quad \text{no-arbitrage constraints}
\end{aligned} \tag{5.1.13}$$

Example 5.5 (Entropy maximization with no-arbitrage constraints and constraints on distribution shape). This problem is similar to (5.1.13). It imposes additional constraints on the distribution to have a bell shape (hump):

$$
\begin{aligned}
&\max_{p_1,\dots,p_n} \quad -\sum_{i=1}^{n} p_i \ln p_i \\
&\text{subject to} \quad \sum_{i=1}^{n} p_i = 1, \quad p_i \geq 0, \quad i = 1,\dots,n, \\
&\left.\begin{aligned}
&\sum_{i=1}^{n} \underline{a}_{ij}\, p_i \leq 0, \quad j = 1,\dots,m, \\
&\sum_{i=1}^{n} \overline{a}_{ij}\, p_i \geq 0, \quad j = 1,\dots,m,
\end{aligned}\right\} \quad \text{no-arbitrage constraints} \\
&1 \leq w_l \leq w_r \leq n, \\
&\frac{p_{i-1} + p_{i+1}}{2} \geq p_i, \quad i = 2,\dots,w_l - 1 \qquad \text{(left slope is convex)} \\
&\frac{p_{i-1} + p_{i+1}}{2} \leq p_i, \quad i = w_l + 1,\dots,w_r - 1 \qquad \text{(hump is concave)} \\
&\frac{p_{i-1} + p_{i+1}}{2} \geq p_i, \quad i = w_r + 1,\dots,n - 1 \qquad \text{(right slope is convex)}
\end{aligned}
\tag{5.1.14}
$$

where w_l and w_r are indices of points of inflection.

In Sect. 9.17, the case study "*Implied Copula CDO Pricing Model: Entropy Approach*" implements the problems (5.1.13) and (5.1.14) in Portfolio Safeguard and solves the problems with real-life data.

5.2 Relative Entropy Minimization

The problem (4.1.4) provides an important insight: maximizing the log-likelihood function in (4.1.4) is equivalent to minimizing the *relative entropy* or *Kullback–Leibler divergence measure*:

$$
D_{\mathrm{KL}}(Y||X) = \sum_{k=1}^{n} q_k \ln \frac{q_k}{p_k} = \sum_{k=1}^{n} q_k \ln q_k - \sum_{k=1}^{n} q_k \ln p_k
$$

with respect to unknown distribution $p = (p_1,\dots,p_l)$ of X given the sample distribution $q = (q_1,\dots,q_l)$ of Y, where $q_k = n_k/n$, $k = 1,\dots,l$.

This observation has far-reaching implications: q should not necessarily be a sample distribution and can be replaced either by a prior probability distribution or by an arbitrary reference probability distribution.

However, $D_{\text{KL}}(Y||X) \neq D_{\text{KL}}(X||Y)$, and the *minimum relative entropy principle (MinEnt)* or *the principle of minimum discrimination information* aims to find a random variable X that minimizes the relative entropy $D_{\text{KL}}(X||Y)$ for a given reference random variable Y subject to any additional constraints on X:

$$\min_{X \in \mathscr{X}} D_{\text{KL}}(X||Y) \quad \text{where } \mathscr{X} \text{ is a feasible set of } X.$$

It is widely used in statistical decision problems dealing with estimation of unknown probability distributions under various constraints. If Y is uniformly distributed, then minimizing $D_{\text{KL}}(X||Y)$ is equivalent to maximizing the Shannon entropy $S(X)$.

Example 5.6 (Relative entropy minimization with linear constraints). The problem of finding a discrete probability distribution $p = (p_1, \ldots, p_n)$ closest to a given probability distribution $q = (q_1, \ldots, q_n)$ in the sense of relative entropy subject to linear constraints on p is formulated by

$$\begin{aligned} \min_{p_1,\ldots,p_n} &\sum_{i=1}^{n} p_i \ln \frac{p_i}{q_i} \\ \text{subject to } &\sum_{i=1}^{n} p_i = 1, \quad Ap \leq b, \\ &l_i \leq p_i \leq u_i, \quad i = 1, \ldots, n, \end{aligned} \tag{5.2.1}$$

where real-valued matrix $A = \{a_{ij}\}_{i,j=1}^{n,m}$ and vector $b \in \mathbb{R}^m$ are known and l_i and u_i are lower and upper bounds such that $0 \leq l_i \leq u_i \leq 1$, $i = 1, \ldots, n$.

5.3 Renyi Entropy Maximization

The Renyi entropy (1.5.4) and (1.5.5) can be used in place of the Shannon entropy (1.5.1) and (1.5.2) in entropy maximization problems. For $\alpha \neq 1$, maximizing the Renyi entropy is equivalent to maximizing $\frac{1}{1-\alpha} \sum_{k=1}^{n} p_k^{\alpha}$ in the discrete case and to maximizing $\frac{1}{1-\alpha} \int_a^b f_X(t)^{\alpha} dt$ in the continuous case. Thus, the entropy maximization problem (5.1.2) with the Renyi entropy for $\alpha \neq 1$ ($\alpha > 0$) is formulated by

$$\max_{p_1,\ldots,p_n} \frac{1}{1-\alpha}\sum_{j=1}^{n} p_j^{\alpha}$$

$$\text{subject to} \quad \sum_{j=1}^{n} x_j^k\, p_j = \mu_k, \quad k = 0, 1, \ldots, m,$$

$$p_j \geq 0, \quad j = 1, \ldots, n, \tag{5.3.1}$$

where all $x_1 \in \mathbb{R}, \ldots, x_n \in \mathbb{R}$ are assumed to be distinct, whereas the entropy maximization problem (5.1.3) is rewritten as

$$\max_{f_X(t)} \frac{1}{1-\alpha}\int_a^b f_X(t)^{\alpha}\, dt$$

$$\text{subject to} \quad \int_a^b t^k\, f_X(t)\, dt = \mu_k, \quad k = 0, 1, \ldots, m,$$

$$f_X(t) \geq 0, \quad a \leq t \leq b, \tag{5.3.2}$$

where a and b are finite ($a < b$) and $f_X(t) \equiv 0$ for $t \notin [a, b]$. In both problems, $\mu_0 = 1$, and $\mu_1 \in \mathbb{R}, \ldots, \mu_m \in \mathbb{R}$ are given. For $\alpha \in (0, 1)$, the multiplier $1/(1-\alpha)$ in the objective functions in (5.3.1) and (5.3.2) can be omitted, whereas for $\alpha > 1$, it is negative and its omission requires changing "max" to "min."

As in the case of Shannon entropy, (5.3.1) is a convex problem with linear constraints, and its objective function is differentiable for any $p_1 > 0, \ldots, p_n > 0$. If the equality (moment) constraints in (5.3.1) hold for some $\tilde{p}_1 > 0, \ldots, \tilde{p}_n > 0$ (Slater's constraint qualification condition), then the optimality conditions for (5.3.1) yield

$$p_j = \left(\sum_{k=0}^{m} \rho_k\, x_j^k\right)^{\frac{1}{\alpha-1}}, \qquad j = 1, \ldots, n, \tag{5.3.3}$$

provided that there exist $\rho_0, \rho_1, \ldots, \rho_m$ such that (5.3.3) satisfies the constraints in (5.3.1). Moreover, if exists, the solution (5.3.3) is the global maximum of (5.3.1); see, e.g., Theorem 3.34 in [51].

Similarly, the necessary optimality conditions for (5.3.2) yield

$$f_X(t) = \left(\sum_{k=0}^{m} \lambda_k\, t^k\right)^{\frac{1}{\alpha-1}} I_{\{t \in [a,b]\}}, \tag{5.3.4}$$

provided that there exist $\lambda_0, \lambda_1, \ldots, \lambda_m$ such that (5.3.4) satisfies the constraints in (5.3.2).

Example 5.7 (No prior information, $m = 0$). Given no prior information about a random variable X either discretely or continuously distributed, the probability distribution of X that maximizes the Renyi entropy is uniform and is given by (5.1.6).

For the collision entropy, i.e., when $\alpha = 2$, determining the maximum entropy distribution reduces to solving a system of linear equations.

Example 5.8 (Collision entropy, $\alpha = 2$). When $\alpha = 2$, the parameters $\rho_0, \rho_1, \ldots, \rho_m$ in (5.3.3) are found from the linear system

$$\sum_{i=0}^{m} \rho_i \left(\sum_{j=1}^{n} x_j^{k+i} \right) = \mu_k, \qquad k = 0, 1, \ldots, m,$$

and if $\sum_{k=0}^{m} \rho_k \, x_j^k \geq 0$ for all $j = 1, \ldots, n$, the solution to (5.3.1) is given by (5.3.3), otherwise, it does not exist. Similarly, $\lambda_0, \lambda_1, \ldots, \lambda_m$ in (5.3.4) are found from the linear system

$$\sum_{i=0}^{m} \frac{b^{k+i+1} - a^{k+i+1}}{k+i+1} \lambda_i = \mu_k, \qquad k = 0, 1, \ldots, m,$$

and if $\sum_{k=0}^{m} \lambda_k \, t^k \geq 0$ for all $t \in [a, b]$, the solution to (5.3.2) is given by (5.3.4), otherwise, it does not exist.

The next example assumes that in (5.3.2), unknown PDF $f_X(t)$ is defined on $\mathbb{R}$ rather than on the finite interval $[a, b]$. In this case, we require that $\int_{-\infty}^{\infty} f_X(t)^\alpha dt < \infty$.

Example 5.9 (Continuously distributed X on $\mathbb{R}$ with $\mu = 0$ and $\sigma = 1$). Let X be a random variable continuously distributed on $\mathbb{R}$ with PDF $f_X(t)$. Then $f_X(t)$ that maximizes the Renyi entropy subject to the constraints $E[X] = 0$ and $\sigma(X) = 1$ exists for $\alpha \in (1/3, 1) \cup (1, \infty)$ and is given by

$$f_X(t) = c \left[1 - \frac{\alpha - 1}{3\alpha - 1} t^2 \right]_+^{\frac{1}{\alpha - 1}}, \qquad t \in \mathbb{R}, \tag{5.3.5}$$

with c defined by

$$c = \begin{cases} \dfrac{\Gamma\left(\frac{1}{1-\alpha}\right)}{\sqrt{\pi}\,\Gamma\left(\frac{1}{1-\alpha} - \frac{1}{2}\right)} \sqrt{\dfrac{1-\alpha}{3\alpha - 1}}, & \alpha < 1, \\[2ex] \dfrac{\Gamma\left(\frac{\alpha}{\alpha-1} + \frac{1}{2}\right)}{\sqrt{\pi}\,\Gamma\left(\frac{\alpha}{\alpha-1}\right)} \sqrt{\dfrac{\alpha - 1}{3\alpha - 1}}, & \alpha > 1, \end{cases}$$

where $[t]_+ = \max\{0, t\}$ and $\Gamma(\cdot)$ is the gamma function; see [9, 24] for details.

5.4 Entropy Maximization with Deviation Measures

How to estimate the probability distribution of a random variable X if its mean and a deviation measure other than the standard deviation are known?

The problem of maximizing the Shannon entropy $S(X)$ for a continuously distributed random variable $X \in \mathscr{L}^1(\Omega)$, whose mean and law-invariant deviation $\mathscr{D} : \mathscr{L}^p(\Omega) \mapsto [0, \infty]$, $p \in [1, \infty]$, are known to be μ and d, respectively, is formulated by

$$\max_{X \in \mathscr{L}^1(\Omega)} S(X) \quad \text{subject to} \quad E[X] = \mu, \quad \mathscr{D}(X) = d. \tag{5.4.1}$$

Let $X_0 \in \mathscr{L}^1(\Omega)$ be a new random variable with a PDF $f_{X_0}(t)$, and let $X = d\, X_0 + \mu$. Then the PDF and the entropy of X are given by

$$f_X(t) = \frac{1}{d} f_{X_0}\left(\frac{t-\mu}{d}\right), \qquad S(X) = S(X_0) + \ln d,$$

respectively, and the problem (5.4.1) simplifies to

$$\max_{X_0 \in \mathscr{L}^1(\Omega)} S(X_0) \quad \text{subject to} \quad E[X_0] = 0, \quad \mathscr{D}(X_0) = 1. \tag{5.4.2}$$

For standard deviation, mean absolute deviation (MAD), lower range deviation, standard lower semideviation, and CVaR deviation, the problem (5.4.2) can be recast in the form (5.1.11), and in these cases, solutions to (5.4.2) are given by (5.1.12):

$\mathscr{D}(X_0)$	Support	$f_{X_0}(t)$	$S(X_0)$
$\sigma(X_0)$	$(-\infty, \infty)$	$\frac{1}{\sqrt{2\pi}} \exp\left(-\frac{1}{2}t^2\right)$	$\frac{1}{2}(1 + \ln[2\pi])$
$\mathrm{MAD}(X_0)$	$(-\infty, \infty)$	$\frac{1}{2}\exp(-\|t\|)$	$1 + \ln 2$
$E[X_0] - \inf X_0$	$[-1, \infty)$	$\exp(-t-1)$	1
$\sigma_-(X_0)$	$(-\infty, \infty)$	$c_1 \exp\left(c_2 t - \frac{1}{2}[-t]_+^2\right)$	1.84434
$\mathrm{CVaR}_\alpha^\Delta(X_0)$	$(-\infty, \infty)$	$(1-\alpha)\exp\left(c_\alpha - t - \frac{1}{\alpha}[c_\alpha - t]_+\right)$	$1 - \ln(1-\alpha)$

where $c_1 \approx 0.260713$, $c_2 \approx -0.638833$, and $c_\alpha = (2\alpha - 1)/(1-\alpha)$; see [16] for details. Figure 5.1 illustrates the function $f_{X_0}(t)$ for $\mathrm{CVaR}_\alpha^\Delta$ with $\alpha = 0.01$, 0.3, 0.5, 0.7, 0.8, and 0.9; see Example 5.11 for obtaining $f_{X_0}(t)$.

A problem closely related to (5.4.1) is maximizing the Shannon entropy for a continuously distributed random variable $Y \in \mathscr{L}^1(\Omega)$ subject to a constraint on the deviation $\mathscr{D}$ projected from a nondegenerate error measure $\mathscr{E}$ and subject to a constraint on the statistic $\mathscr{S}$ associated with $\mathscr{E}$:

$$\max_{Y \in \mathscr{L}^1(\Omega)} S(Y) \quad \text{subject to} \quad \mathscr{D}(Y) = d, \quad c \in \mathscr{S}(Y), \tag{5.4.3}$$

where $\mathscr{D}$ and $\mathscr{S}$ are defined by (2.4.3) and (2.4.4), respectively.

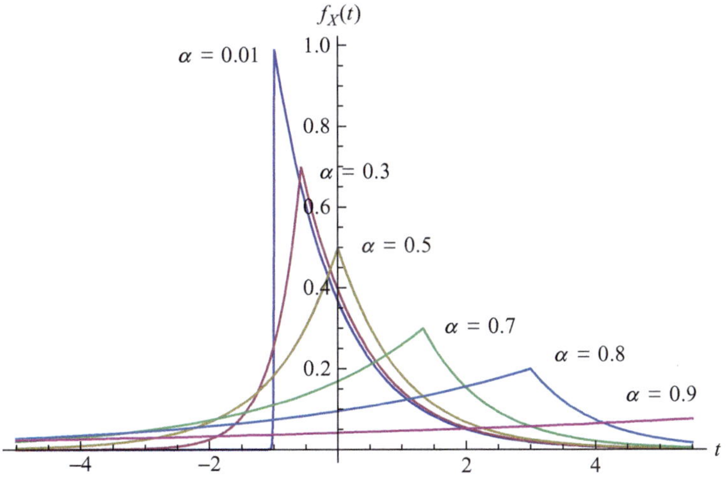

Fig. 5.1 The PDF $f_X(t)$ that maximizes the Shannon entropy $S(X)$ subject to constraints on the mean and CVaR deviation: $E[X]=0$ and $\text{CVaR}_\alpha^\Delta(X)=1$ for $\alpha=0.01$, 0.3, 0.5, 0.7, 0.8, and 0.9

Let $Y_0 \in \mathscr{L}^1(\Omega)$ be a new random variable with a PDF $f_{Y_0}(t)$, and let $Y = d\, Y_0 + c$. Then the PDF and the entropy of Y are given by

$$f_Y(t) = \frac{1}{d} f_{Y_0}\left(\frac{t-c}{d}\right), \qquad S(Y) = S(Y_0) + \ln d,$$

respectively, and the problem (5.4.3) simplifies to

$$\max_{Y_0 \in \mathscr{L}^1(\Omega)} S(Y_0) \qquad \text{subject to} \quad \mathscr{D}(Y_0) = 1, \quad 0 \in \mathscr{S}(Y_0). \tag{5.4.4}$$

Proposition 5.1. *Let $Z^* \in \mathscr{L}^1(\Omega)$ be a continuously distributed random variable that maximizes the Shannon entropy subject to a constraint on a nondegenerate error measure $\mathscr{E}$:*

$$\max_{Z \in \mathscr{L}^1(\Omega)} S(Z) \qquad \textit{subject to} \quad \mathscr{E}(Z) = 1. \tag{5.4.5}$$

If the deviations $\mathscr{D}$ in (5.4.1) *and in* (5.4.3) *are projected from $\mathscr{E}$, then the random variables X_0^* and Y_0^* that solve the problems* (5.4.2) *and* (5.4.4)*, respectively, are determined by*

$$X_0^* = Z^* - E[Z^*], \qquad Y_0^* = Z^* - C^*, \quad C^* \in \mathscr{S}(Z^*), \tag{5.4.6}$$

where $\mathscr{S}$ is the statistic associated with $\mathscr{E}$ and

$$-E[Y_0^*] \in \mathscr{S}(X_0^*).$$

Proof. Let X_0 and Y_0 be feasible random variables in (5.4.2) and (5.4.4), respectively, and let $C_0 \in \mathscr{S}(X_0)$. Then (5.4.2) reduces to (5.4.4) by substitution $Y_0 = X_0 - C_0$. Indeed, $S(X_0) = S(Y_0)$, $0 \in \mathscr{S}(Y_0)$, and $\mathscr{D}(X_0) = \mathscr{D}(Y_0)$.

Now, since $\mathscr{D}$ is projected from $\mathscr{E}$, the constraint $0 \in \mathscr{S}(Y_0)$ implies that $\mathscr{D}(Y_0) = \mathscr{E}(Y_0 - 0) = \mathscr{E}(Y_0)$ and the problem (5.4.4) can be equivalently restated as

$$\max_{Y_0 \in \mathscr{L}^1(\Omega)} S(Y_0) \qquad \text{subject to} \quad \mathscr{E}(Y_0) = 1, \quad 0 \in \mathscr{S}(Y_0). \tag{5.4.7}$$

Let Z^* be a solution to (5.4.5), let $C^* \in \mathscr{S}(Z^*)$, and let $Y_0^* = Z^* - C^*$, then $\mathscr{E}(Y_0^*) = 1$. Indeed, by definition of the statistic $\mathscr{S}$ associated with $\mathscr{E}$,

$$\mathscr{E}(Y_0^*) = \mathscr{E}(Z^* - C^*) = \min_{C \in \mathbb{R}} \mathscr{E}(Z^* - C) \le \mathscr{E}(Z^*) = 1,$$

so that $\mathscr{E}(Y_0^*) \le 1$. By contradiction, let $\mathscr{E}(Y_0^*) = \delta < 1$, and let $\tilde{Z} = Y_0^*/\delta$. Then, the positive homogeneity of $\mathscr{E}$ implies that $\mathscr{E}(\tilde{Z}) = 1$ and $S(\tilde{Z}) = S(Y_0^*) - \ln\delta = S(Z^*) - \ln\delta > S(Z^*)$, so that Z^* is not optimal for (5.4.5). Thus, $\mathscr{E}(Y_0^*) = 1$.

Now, compared to (5.4.5), the problem (5.4.7) has the additional constraint $0 \in \mathscr{S}(Y_0)$, and consequently, its optimal value is less than or equal to that of (5.4.5), i.e., $S(Y_0) \le S(Z^*)$ for any feasible Y_0. However, $0 \in \mathscr{S}(Y_0^*)$ and $\mathscr{E}(Y_0^*) = 1$, so that Y_0^* is feasible for (5.4.7), and since $S(Y_0^*) = S(Z^*)$, Y_0^* is also optimal for (5.4.7).

Finally, the constraint $E[X_0] = 0$ and the relationship $Y_0 = X_0 - C_0$ with $C_0 \in \mathscr{S}(X_0)$ imply that $E[Y_0] = E[X_0] - C_0 = -C_0$. Consequently, $-E[Y_0] \in \mathscr{S}(X_0)$ and $X_0 = Y_0 + C_0 = Y_0 - E[Y_0]$, and optimal X_0^* in (5.4.2) is determined by $X_0^* = Y_0^* - E[Y_0^*] = Z^* - C^* - E[Z^* - C^*] = Z^* - E[Z^*]$. □

For the error measure (2.4.1): $\mathscr{E}(Z) = \|a\,Z_+ + b\,Z_-\|_p$ with $a > 0$, $b > 0$, and $p \in [1, \infty)$, the problem (5.4.5) can be represented in the form (5.1.11) with $m = 1$, $V = (-\infty, \infty)$, and

$$h_1(t) = (a\,\max\{t, 0\} + b\,\max\{-t, 0\})^p.$$

Its solution is given by (5.1.12), where c and λ_1 are found from the constraints $\int_{-\infty}^{\infty} f_Z(t)\,dt = 1$ and $\int_{-\infty}^{\infty} h_1(t) f_Z(t)\,dt = 1$, so that the PDF of optimal Z^* in (5.4.5) is determined by

$$f_{Z^*}(t) = \frac{\exp\left(-\frac{1}{p}(a\,\max\{t,0\} + b\,\max\{-t,0\})^p\right)}{(a^{-p} + b^{-p})\,p^{1/p}\,\Gamma[(p+1)/p]}, \qquad t \in \mathbb{R}, \tag{5.4.8}$$

and

$$S(Z^*) = \frac{1 + \ln p}{p} + \ln\left((a^{-p} + b^{-p})\,\Gamma\left[\frac{p+1}{p}\right]\right).$$

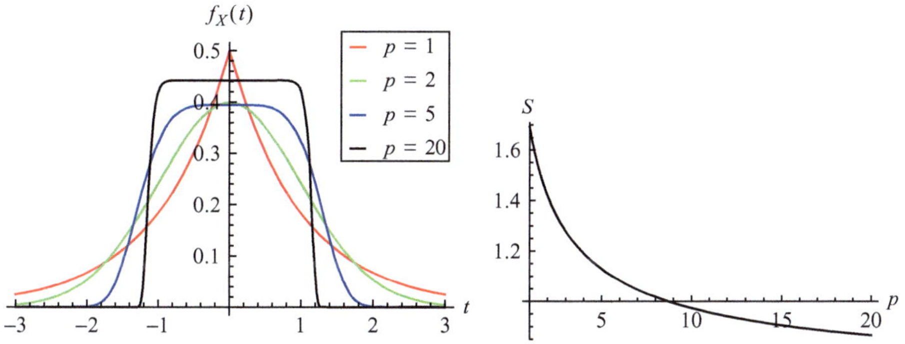

Fig. 5.2 PDFs $f_Z(t)$ that maximize the Shannon entropy $S(Z)$ subject to the constraint $\|Z\|_p = 1$ for $p = 1, 2, 5, 20$ and the value of the entropy $S(Z)$ for these PDFs as a function of $p \in [1, \infty)$. The PDF with the sharp spike corresponds to $p = 1$, and the PDF dispersion decreases with p

Example 5.10 (Entropy maximization with a constraint on p-norm). For $a = b = 1$, the error measure (2.4.1) simplifies to $\|X\|_p$, and consequently, the PDF of a random variable Z^* that solves the problem (5.4.5) with $\mathscr{E}(Z) = \|Z\|_p$ is given by (5.4.8) with $a = b = 1$:

$$f_{Z^*}(t) = \frac{1}{2p^{1/p}\Gamma[(p+1)/p]} \exp\left(-\frac{|t|^p}{p}\right), \qquad t \in \mathbb{R}. \tag{5.4.9}$$

Figure 5.2 shows the function (5.4.9) for $p = 1, 2, 5$, and 20 and also depicts $S(Z^*) = (1 + \ln p)/p + \ln(2\Gamma[(p+1)/p])$ as a function of $p \in [1, \infty)$. Remarkably, $\lim_{p\to\infty} f_{Z^*}(t) = \frac{1}{2} I_{\{-1 \le t \le 1\}}$ and $\lim_{p\to\infty} S(Z^*) = \ln 2$.

In this case, $E[Z^*] = 0$, and (5.4.6) implies that X_0^* solving (5.4.2) is given by $X_0^* = Z^*$, so that $f_{X_0^*}(t) = f_{Z^*}(t)$. Since σ and MAD are projected from the error measures $\|\cdot\|_2$ and $\|\cdot\|_1$, respectively, the PDFs of optimal X_0^* in (5.4.2) with $\mathscr{D} = \sigma$ and $\mathscr{D} =$ MAD are given by (5.4.9) for $p = 2$ and $p = 1$, respectively.

Example 5.11 (Entropy maximization with a constraint on asymmetric mean absolute error). For $p = 1, a = 1$, and $b = 1/\alpha - 1$, the error measure (2.4.1) reduces to (2.0.3), and consequently, the PDF of a random variable Z^* that solves the problem (5.4.5) with $\mathscr{E}(Z) = \mathscr{E}_\alpha(Z)$ is given by (5.4.8) with $p = 1, a = 1$, and $b = 1/\alpha - 1$:

$$f_{Z^*}(t) = (1-\alpha)\exp\left(\frac{1}{\alpha}\min\{t, 0\} - t\right), \qquad t \in \mathbb{R}, \tag{5.4.10}$$

for which

$$E[Z^*] = -\frac{2\alpha - 1}{1-\alpha}, \qquad q_{Z^*}^+(\alpha) = 0, \qquad S(Z^*) = 1 - \ln(1-\alpha).$$

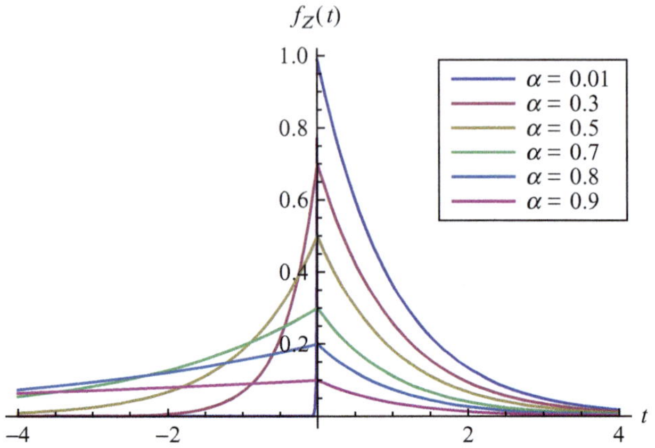

Fig. 5.3 The PDF $f_Z(t)$ that maximizes the Shannon entropy $S(Z)$ subject to a constraint on the asymmetric mean absolute error (2.0.3): $\mathscr{E}_\alpha(Z) = 1$ for $\alpha = 0.01, 0.3, 0.5, 0.7, 0.8$, and 0.9

Figure 5.3 illustrates the function $f_{Z^*}(t)$ for $\alpha = 0.01, 0.3, 0.5, 0.7, 0.8$, and 0.9.

The deviation $\mathscr{D}$ projected from $\mathscr{E}_\alpha$ and the statistic $\mathscr{S}$ associated with $\mathscr{E}_\alpha$ are CVaR deviation and α-quantile, respectively. Let X_0^* solve (5.4.2) with $\mathscr{D} = \text{CVaR}_\alpha^\Delta$:

$$\max_{X_0 \in \mathscr{L}^1(\Omega)} S(X_0) \qquad \text{subject to} \quad E[X_0] = 0, \quad \text{CVaR}_\alpha^\Delta(X_0) = 1,$$

and let Y_0^* solve (5.4.4) with $\mathscr{D} = \text{CVaR}_\alpha^\Delta$ and $\mathscr{S}(Y_0) = q_{Y_0}(\alpha)$:

$$\max_{Y_0 \in \mathscr{L}^1(\Omega)} S(Y_0) \qquad \text{subject to} \quad \text{CVaR}_\alpha^\Delta(Y_0) = 1, \quad q_{Y_0}^+(\alpha) = 0.$$

Then the relationships (5.4.6) imply that

$$X_0^* = Z^* + c_\alpha, \qquad Y_0^* = Z^*,$$

where $c_\alpha = (2\alpha - 1)/(1 - \alpha)$, and the PDF of X_0^* is determined by

$$f_{X_0^*}(t) = (1 - \alpha) \exp\left(\frac{1}{\alpha} \min\{t - c_\alpha, 0\} - t + c_\alpha\right), \qquad t \in \mathbb{R} \tag{5.4.11}$$

(see Fig. 5.1), whereas the PDF of Y_0^* coincides with (5.4.10) (see Fig. 5.3).

Another approach to entropy maximization is based on a quantile representation of the entropy. For a continuously distributed X with a CDF $F_X(t)$, the Shannon entropy $S(X)$ can be represented by

$$S(X) = \int_0^1 \ln q'_X(s)\, ds, \qquad q_X(s) = F_X^{-1}(s),$$

where $q'_X(s)$ is the derivative of the s-quantile $q_X(s)$ with respect to s. (For continuous $F_X(t)$, $q_X^+(s) = q_X^-(s)$, so there is no ambiguity in using $q_X(s)$.) Then, for example, for the mixed-CVaR deviation in the form (2.1.6), the problem (5.4.2) can be formulated as

$$\max_{q_{X_0}(s)} \int_0^1 \ln q'_{X_0}(s)\, ds \quad \text{subject to} \quad \int_0^1 q_{X_0}(s)\, ds = 0, \quad \int_0^1 q_{X_0}(s)\, \phi(s)\, ds = -1,$$

where $\phi(s)$ is a given left-continuous and nonincreasing function with $\phi(0+) < \infty$, $\phi(1-) = 0$, and $\int_0^1 \phi(s)ds = 1$. This *isoperimetric problem of calculus of variations* is solved in [16].

Other examples of distributions that maximize either Shannon or Renyi differential entropy subject to constraints on the mean and deviation are discussed in [16]. The following general results hold [16]:

(i) A solution $X \in \mathscr{L}^1(\Omega)$ to (5.4.1) with an arbitrary law-invariant deviation measure has a log-concave PDF, i.e., $\ln f_X(t)$ is concave.
(ii) For any $X \in \mathscr{L}^1(\Omega)$ with a log-concave PDF $f_X(t)$, there exists a comonotone deviation measure $\mathscr{D}$[1] such that X is optimal in (5.4.1) with $\mathscr{D}$. In other words, the *inverse maximum entropy problem* for (5.4.1): finding $\mathscr{D}$ for a given log-concave PDF $f_X(t)$, has a solution in the class of comonotone deviations; see [16].

5.5 Entropy Maximization with Constraints on Average α-Quantile for Different Confidence Levels

The problem (5.1.1) finds the maximum entropy distribution of a random variable X subject to constraints on the first m moments of X. Suppose that the expected value of $X \in \mathscr{L}^1(\Omega)$ is given to be μ (first moment), but instead of knowing the higher moments, we assume that the average α-quantiles of X with confidence levels $\alpha_1 < \ldots < \alpha_m < \alpha_{m+1} = 1$ are real-valued $\mu_1, \ldots, \mu_m, \mu_{m+1} = \mu$, respectively ($\alpha_1 > 0$ and the constraint $E[X] = \mu$ is reformulated as $\overline{q}_X(1) = \mu_{m+1}$). In this case, the PDF of X that maximizes the Shannon entropy can be found from the problem similar to (5.1.1):

[1] $\mathscr{D}$ is *comonotone* if $\mathscr{D}(X + Y) = \mathscr{D}(X) + \mathscr{D}(Y)$ for any two comonotone random variables $X \in \mathscr{L}^p(\Omega)$ and $Y \in \mathscr{L}^p(\Omega)$, whereas the random variables X and Y are comonotone, if there exists a set $A \subseteq \Omega$ such that $\mathbb{P}[A] = 1$ and $(X(\omega_1) - X(\omega_2))(Y(\omega_1) - Y(\omega_2)) \geq 0$ for all $\omega_1, \omega_2 \in A$.

$$\max_{X\in\mathscr{L}^1(\Omega)} \quad S(X) \qquad \text{subject to} \quad \overline{q}_X(\alpha_k) = \mu_k, \quad k = 1,\dots,m+1. \tag{5.5.1}$$

For $m = 1$, (5.5.1) is equivalent to (5.4.1) with CVaR deviation.

Proposition 5.2 (Maximum entropy distribution with given average α-quantiles). *The PDF of a random variable X that solves the problem* (5.5.1) *is given by*

$$f_X(t) = \begin{cases} \dfrac{\alpha_1}{a_1-\mu_1}\exp\left(\dfrac{t-a_1}{a_1-\mu_1}\right), & t\in(-\infty,a_1], \\[2ex] \dfrac{(\alpha_k-\alpha_{k-1})\lambda_k}{e^{\lambda_k(a_k-a_{k-1})}-1}\exp\left(\lambda_k(t-a_{k-1})\right), & t\in(a_{k-1},a_k], k=2,\dots,m, \\[2ex] \dfrac{1-\alpha_m}{\frac{\mu-\alpha_m\mu_m}{1-\alpha_m}-a_m}\exp\left(-\dfrac{t-a_m}{\frac{\mu-\alpha_m\mu_m}{1-\alpha_m}-a_m}\right), & t\in(a_m,\infty), \end{cases} \tag{5.5.2}$$

provided that there exist real-valued $a_1,\dots,a_m$ and $\lambda_2,\dots,\lambda_m$ such that satisfy the equations

$$\frac{a_k-a_{k-1}}{e^{\lambda_k(a_k-a_{k-1})}-1} - \frac{1}{\lambda_k} = \frac{\alpha_k\mu_k-\alpha_{k-1}\mu_{k-1}}{\alpha_k-\alpha_{k-1}} - a_k, \quad k=2,\dots,m \tag{5.5.3}$$

and make $f_X(t)$ continuous at $a_1,\dots,a_m$:

$$f_X(a_k-) = f_X(a_k+), \qquad k=1,\dots,m. \tag{5.5.4}$$

Proof. Let $a_1,\dots,a_m$ be unknown α-quantiles of a continuously distributed random variable X corresponding to confidence levels $\alpha_1,\dots,\alpha_m$, and let $a_0 = -\infty$, $a_{m+1} = \infty$, $\alpha_0 = 0$, and μ_0 be any finite number, then (5.5.1) can be reformulated in the form

$$\begin{aligned} &\max_{\substack{f_X(t)\ge 0\\ a_1,\dots,a_m}} \quad -\int_{-\infty}^{\infty} f_X(t)\ln f_X(t)\,dt \\ &\text{subject to} \quad \int_{a_{k-1}}^{a_k} f_X(t)\,dt = \alpha_k-\alpha_{k-1}, \quad k=2,\dots,m, \\ &\qquad\qquad\quad \int_{a_{k-1}}^{a_k} t\,f_X(t)\,dt = \alpha_k\mu_k-\alpha_{k-1}\mu_{k-1}, \qquad k=2,\dots,m, \end{aligned} \tag{5.5.5}$$

which yields $m + 1$ independent subproblems:

$$\max_{f_k(t)\geq 0} \; -\int_{a_{k-1}}^{a_k} f_k(t)\ln f_k(t)\,dt$$

$$\text{subject to} \int_{a_{k-1}}^{a_k} f_k(t)\,dt = \alpha_k - \alpha_{k-1}, \qquad \int_{a_{k-1}}^{a_k} t\, f_k(t)\,dt = \alpha_k\mu_k - \alpha_{k-1}\mu_{k-1} \tag{5.5.6}$$

for $k = 1,\ldots,m+1$, where

$$f_X(t) = f_k(t), \qquad t \in (a_{k-1}, a_k], \quad k = 1,\ldots,m+1.$$

Similarly to Boltzmann's theorem [10, Theorem 12.1.1], a solution to (5.5.6) is given by

$$f_k(t) = c_k e^{\lambda_k t}, \qquad k = 1,\ldots,m+1,$$

with c_k and λ_k found from the constraints in (5.5.6), so that $f_X(t)$ is determined by (5.5.2) and $\lambda_2,\ldots,\lambda_m$ should satisfy (5.5.3).

Thus, for (5.5.2), we obtain

$$\begin{aligned} g(a_1,\ldots,a_m) &= -\int_{-\infty}^{\infty} f_X(t)\ln f_X(t)\,dt = -\sum_{k=1}^{m+1}\int_{a_{k-1}}^{a_k} f_k(t)\ln f_k(t)\,dt \\ &= -\alpha_1\left(\ln\frac{\alpha_1}{a_1-\mu_1}-1\right) - (1-\alpha_m)\left(\ln\frac{1-\alpha_m}{\frac{\mu-\alpha_m\mu_m}{1-\alpha_m}-a_m}-1\right) \\ &\quad -\sum_{k=2}^{n}\left[(\alpha_k-\alpha_{k-1})\ln\frac{(\alpha_k-\alpha_{k-1})\lambda_k}{e^{\lambda_k a_k}-e^{\lambda_k a_{k-1}}}+(\alpha_k\mu_k-\alpha_{k-1}\mu_{k-1})\lambda_k\right], \end{aligned}$$

and optimal $a_1,\ldots,a_m$ are found from the conditions $\partial g/\partial a_k = 0$, $k = 1,\ldots,m$. Observe that in view of (5.5.3), $\lambda_k = \lambda_k(a_{k-1}, a_k)$, so that the derivative $\partial g/\partial a_k$ includes terms $(\partial g/\partial\lambda_k)\cdot(\partial\lambda_k/\partial a_k)$ and $(\partial g/\partial\lambda_{k+1})\cdot(\partial\lambda_{k+1}/\partial a_k)$. However, (5.5.3) implies that

$$\frac{\partial g}{\partial\lambda_k} = -\frac{\partial}{\partial\lambda_k}\left[(\alpha_k-\alpha_{k-1})\ln\frac{(\alpha_k-\alpha_{k-1})\lambda_k}{e^{\lambda_k a_k}-e^{\lambda_k a_{k-1}}} + (\alpha_k\mu_k-\alpha_{k-1}\mu_{k-1})\lambda_k\right] = 0.$$

This means that in differentiating $g(a_1,\ldots,a_m)$ with respect to $a_1,\ldots,a_m$, parameters $\lambda_2,\ldots,\lambda_m$ can be treated as constants. Consequently, for $m \geq 2$, we obtain

$$\begin{cases} \dfrac{\partial g}{\partial a_1} = \dfrac{\alpha_1}{a_1 - \mu_1} - \dfrac{(\alpha_2 - \alpha_1)\lambda_2}{e^{\lambda_2(a_2 - a_1)} - 1} = 0, \\ \dfrac{\partial g}{\partial a_k} = \dfrac{(\alpha_k - \alpha_{k-1})\lambda_k e^{\lambda_k(a_k - a_{k-1})}}{e^{\lambda_k(a_k - a_{k-1})} - 1} - \dfrac{(\alpha_{k+1} - \alpha_k)\lambda_{k+1}}{e^{\lambda_{k+1}(a_{k+1} - a_k)} - 1} = 0, \\ \qquad k = 2, \ldots, m-1, \\ \dfrac{\partial g}{\partial a_m} = \dfrac{(\alpha_n - \alpha_{n-1})\lambda_n e^{\lambda_n(a_n - a_{n-1})}}{e^{\lambda_n(a_n - a_{n-1})} - 1} - \dfrac{1 - \alpha_m}{\frac{\mu - \alpha_m \mu_m}{1 - \alpha_m} - a_m} = 0, \end{cases} \tag{5.5.7}$$

where for $m = 2$, the middle equation is omitted. This system is equivalent to the conditions (5.5.4) on continuity of $f_X(t)$ at the points $a_1, \ldots, a_m$. □

If $m = 1$, then (5.5.3) is omitted, and for $\mu = 0$, $\alpha_1 = \alpha$, and $\mu_1 = -1$, (5.5.2) simplifies to (5.4.11).

For $m = 2$, (5.5.3) and (5.5.4) reduce to a system of three equations:

$$\begin{cases} \dfrac{a_2 - a_1}{e^{\lambda_2(a_2 - a_1)} - 1} - \dfrac{1}{\lambda_2} + a_2 = \dfrac{\alpha_2\mu_2 - \alpha_1\mu_1}{\alpha_2 - \alpha_1}, \\ \dfrac{(\alpha_2 - \alpha_1)\lambda_2}{e^{\lambda_2(a_2 - a_1)} - 1} = \dfrac{\alpha_1}{a_1 - \mu_1}, \\ \dfrac{(\alpha_2 - \alpha_1)\lambda_2 e^{\lambda_2(a_2 - a_1)}}{e^{\lambda_2(a_2 - a_1)} - 1} = \dfrac{1 - \alpha_2}{\frac{\mu - \alpha_2\mu_2}{1 - \alpha_2} - a_2}, \end{cases}$$

which implies that

$$\lambda_2 = \frac{1}{\alpha_2 - \alpha_1}\left(\frac{1 - \alpha_2}{\frac{\mu - \alpha_2\mu_2}{1 - \alpha_2} - a_2} - \frac{\alpha_1}{a_1 - \mu_1}\right)$$

and that a_1 and a_2 satisfy

$$\begin{cases} \dfrac{\alpha_1}{a_1 - \mu_1}\left(\dfrac{\alpha_2\mu_2 - \alpha_1\mu_1}{\alpha_2 - \alpha_1} - a_1\right) = \dfrac{1 - \alpha_2}{\frac{\mu - \alpha_2\mu_2}{1 - \alpha_2} - a_2}\left(\dfrac{\alpha_2\mu_2 - \alpha_1\mu_1}{\alpha_2 - \alpha_1} - a_2\right), \\ \dfrac{a_2 - a_1}{\alpha_2 - \alpha_1}\left(\dfrac{1 - \alpha_2}{\frac{\mu - \alpha_2\mu_2}{1 - \alpha_2} - a_2} - \dfrac{\alpha_1}{a_1 - \mu_1}\right) = \ln\left[\dfrac{1 - \alpha_2}{\frac{\mu - \alpha_2\mu_2}{1 - \alpha_2} - a_2} \dfrac{a_1 - \mu_1}{\alpha_1}\right]. \end{cases}$$

The last system can be reduced further just to a single equation with respect to either a_1 or a_2, which then can be solved numerically.

Chapter 6
Regression Models

In statistics, regression analysis aims to find the best relationship between a *response* random variable Y (*regressant*) and n independent variables $x_1, \ldots, x_n$ (*regressors*) in the form

$$Y = f(x_1, \ldots, x_n) + \epsilon,$$

based on m available simultaneous observations of $x_1, \ldots, x_n$ and Y (regression data), $x_{1j}, \ldots, x_{nj}, y_j$, $j = 1, \ldots, m$, where ϵ is the *approximation error*.

There are two main classes of regression: *parametric* and *nonparametric*. If the function f is determined by a finite set of parameters, regression is called parametric; otherwise it is called nonparametric. The class of parametric regressions is further divided into *linear* and *nonlinear*. In linear regression, f is a linear function with respect to unknown parameters, whereas $x_1, \ldots, x_n$ can be involved nonlinearly. Typically, linear regression is assumed to be linear with respect to $x_1, \ldots, x_n$ as well (see, e.g., [19]), and f has the form

$$f(x_1, \ldots, x_n) = c_0 + c_1 x_1 + \ldots + c_n x_n = c_0 + \sum_{k=1}^{n} c_k x_k, \tag{6.0.1}$$

where $c_k \in \mathbb{R}$, $k = 0, 1, \ldots, n$, are unknown regression parameters with c_0 called *intercept* or *bias*. (Estimates of $c_0, c_1, \ldots, c_n$ found from the regression data are denoted by $\hat{c}_0, \hat{c}_1, \ldots, \hat{c}_n$, respectively.) In nonlinear regression, f is a nonlinear function of specified unknown parameters, which are usually found iteratively.

One of the main approaches for finding estimates of regression parameters is to maximize the likelihood of the observations of $y_1, \ldots, y_m$ under the assumption that the residuals $e_j = y_j - f(x_{1j}, \ldots, x_{nj})$, $j = 1, \ldots, m$, are realizations of *independent and identically distributed* (iid) random variables $\epsilon_1, \ldots, \epsilon_m$ with zero mean; see [60]. For example, if $\epsilon_1, \ldots, \epsilon_m$ are iid and have the normal distribution $N\left(0, \sigma^2\right)$, then the likelihood of observing $y_1, \ldots, y_m$ is given by

M. Zabarankin and S. Uryasev, *Statistical Decision Problems: Selected Concepts and Portfolio Safeguard Case Studies*, Springer Optimization and Its Applications 85, DOI 10.1007/978-1-4614-8471-4_6, © Springer Science+Business Media New York 2014

$$\frac{1}{(\sqrt{2\pi}\sigma)^m}\prod_{j=1}^{m}\exp\left(-\frac{1}{2\sigma^2}\left(y_j - f(x_{1j},\ldots,x_{nj})\right)^2\right),$$

and its maximization simplifies to

$$\min\sum_{j=1}^{m}\left(y_j - f(x_{1j},\ldots,x_{nj})\right)^2,$$

which is called *least squares problem*. With $f(x_1,\ldots,x_n)$ in the form (6.0.1), this problem yields a system of linear equations for estimates $\hat{c}_1,\ldots,\hat{c}_n$:

$$\sum_{k=1}^{n}\hat{c}_k\sum_{j=1}^{n}\left(x_{ij}-\tilde{x}_i\right)\left(x_{kj}-\tilde{x}_k\right) = \sum_{j=1}^{m}\left(x_{ij}-\tilde{x}_i\right)\left(y_j-\tilde{y}\right), \quad i = 1,\ldots,n, \tag{6.0.2}$$

with $\hat{c}_0 = \tilde{y} - \sum_{k=1}^{n} c_k\tilde{x}_k$, where $\tilde{x}_i = \frac{1}{n}\sum_{j=1}^{m} x_{ij}$ for $i = 1,\ldots,n$ and $\tilde{y} = \frac{1}{n}\sum_{j=1}^{m} y_j$.

Even if $\epsilon_1,\ldots,\epsilon_m$ are only uncorrelated (not necessarily independent) with zero mean and same variance, the *Gauss–Markov theorem* states that the *best linear unbiased estimator* (BLUE) of the form (6.0.1) is determined by *least squares linear regression*. If $\epsilon_1,\ldots,\epsilon_m$ are correlated and/or not identically distributed random variables, then least squares regression may not be appropriate.

Statistical approximation theory takes a different perspective on regression: when the response random variable Y is not understood completely and is better to be treated as a function $f(X_1,\ldots,X_n)$ of *random* variables $X_1,\ldots,X_n$, the error $Y - f(X_1,\ldots,X_n)$ is sought to minimize some loss function or *error measure* with respect to unknown regression parameters; see [46]. In this approach, central to regression analysis is the choice of error measure that should conform to *risk preferences* of an analyst. For example, if the problem is to track a stock market index by a portfolio of selected financial instruments, whose returns are random variables $X_1,\ldots,X_n$, the analyst may penalize only underperformance of the portfolio return $f(X_1,\ldots,X_n)$ with respect to the index return Y, so that symmetric measures like $\|\cdot\|_2$ are not appropriate.

This chapter pursues the statistical approximation approach to regression. It focuses on a general theory of approximating an output random variable Y by a linear combination of input random variables $X_1,\ldots,X_n$:

$$f(X_1,\ldots,X_n) = c_0 + c_1X_1 + \ldots + c_nX_n = c_0 + \sum_{k=1}^{n} c_kX_k$$

with an arbitrary error measure $\mathscr{E}$ under additional constraints on regression coefficients.

6.1 Generalized Linear Regression and Error Decomposition

An unconstrained generalized linear regression problem is formulated as follows: *approximate a random variable* $Y \in \mathscr{L}^2(\Omega)$ *by a linear combination* $c_0 + \sum_{k=1}^n c_k X_k$ *of given random variables* $X_k \in \mathscr{L}^2(\Omega)$, $k = 1, \ldots, n$, *and minimize an error measure* $\mathscr{E}$ *of the error* $Z = Y - c_0 - \sum_{k=1}^n c_k X_k$ *with respect to real-valued* $c_0, c_1, \ldots, c_n$, *where* $\mathscr{E}$ *is assumed to be nondegenerate and finite everywhere on* $\mathscr{L}^2(\Omega)$, or, formally,

$$\min_{c_0, c_1, \ldots, c_n} \mathscr{E}(Z) \quad \text{with} \quad Z = Y - c_0 - \sum_{k=1}^n c_k X_k. \tag{6.1.1}$$

Observe that because of possible asymmetry of $\mathscr{E}$, $\mathscr{E}(-Z) \neq \mathscr{E}(Z)$.

Well-known particular cases of the linear regression (6.1.1) include

(a) Least squares linear regression ($\mathscr{E}(Z) = \|Z\|_2$)
(b) Median regression ($\mathscr{E}(Z) = \|Z\|_1$)
(c) Quantile regression ($\mathscr{E}(Z)$ is the asymmetric mean absolute error (2.0.3))

The choice of error measure to be used in a given regression problem is determined by particular application and risk preferences of a decision maker.

Theorem 3.2 in [46] shows that if $\mathscr{D}$ is the deviation projected from $\mathscr{E}$ (defined by (2.4.2)) and $\mathscr{S}$ is the statistic associated with $\mathscr{E}$ (defined by (2.4.4)), then an optimal solution $c_0^*, c_1^*, \ldots, c_n^*$ of the generalized linear regression (6.1.1) can be found in two stages: (i) minimizing $\mathscr{D}\left(Y - \sum_{k=1}^n c_k X_k\right)$ with respect to $c_1, \ldots, c_n$, i.e. $(c_1^*, \ldots, c_n^*) \in \arg\min_{c_1, \ldots, c_n} \mathscr{D}\left(Y - \sum_{k=1}^n c_k X_k\right)$, and (ii) setting the intercept c_0^* to any value of $\mathscr{S}\left(Y - \sum_{k=1}^n c_k^* X_k\right)$. In other words, (6.1.1) is reduced to

$$(c_1^*, \ldots, c_n^*) \in \arg\min_{c_1, \ldots, c_n} \mathscr{D}\left(Y - \sum_{k=1}^n c_k X_k\right) \quad \text{and} \quad c_0^* \in \mathscr{S}\left(Y - \sum_{k=1}^n c_k^* X_k\right). \tag{6.1.2}$$

This result is known as *error decomposition*.

Further, Theorem 4.1 in [46] states that $c_1^*, \ldots, c_n^*$ solve (6.1.2) if and only if

$$\text{there exists } Q \in \mathscr{Q}_{\mathscr{D}}\left(Y - \sum_{k=1}^n c_k^* X_k\right) \text{ such that } E[(1 - Q)X_j] = 0$$
$$\text{for } j = 1, \ldots, n, \tag{6.1.3}$$

where $\mathscr{Q}_{\mathscr{D}}\left(Y - \sum_{k=1}^n c_k^* X_k\right)$ is the set of risk identifiers for $Y - \sum_{k=1}^n c_k^* X_k$ with respect to the deviation measure $\mathscr{D}$ (see Sect. 2.2).

In many applications, e.g., factor models, index tracking, and replication problems, the coefficients $c_0, c_1, \ldots, c_n$ are often required to satisfy additional constraints. Let $\mathscr{C} \subseteq \mathbb{R}^{n+1}$ be a feasible set of $(n + 1)$-dimensional vector

$c = (c_0, c_1, \ldots, c_n)$. For example, the requirement of $c_0, c_1, \ldots, c_n$ to be nonnegative translates into having $\mathscr{C} = \{c \in \mathbb{R}^{n+1} \,|\, c \geq 0\}$. A constrained generalized linear regression takes the form

$$\min_{c_0, c_1, \ldots, c_n} \mathscr{E}\left(Y - c_0 - \sum_{k=1}^{n} c_k X_k\right) \quad \text{subject to} \quad (c_0, c_1, \ldots, c_n) \in \mathscr{C}. \tag{6.1.4}$$

Next sections discuss the problem (6.1.4) with different error measures $\mathscr{E}$ and with typical feasible sets $\mathscr{C}$ arising in various statistical decision applications.

6.2 Least Squares Linear Regression

A least squares linear regression is one of the basic and most widely used statistical tools that finds its applications in virtually all areas of science dealing with data analysis and statistics, e.g., physics, biology, medicine, finance, and economics.

6.2.1 Unconstrained Least Squares Linear Regression

Unconstrained least squares linear regression is a particular case of (6.1.1) with $\mathscr{E}(\cdot) = \|\cdot\|_2$ and is given by

$$\min_{c_0, c_1, \ldots, c_n} \left\| Y - c_0 - \sum_{k=1}^{n} c_k X_k \right\|_2^2. \tag{6.2.1}$$

The first-order necessary optimality conditions for the optimization problem (6.2.1) yield a system of linear equations for $c_0^*, c_1^*, \ldots, c_n^*$:

$$\begin{cases} \displaystyle\sum_{k=1}^{n} c_k^* \operatorname{Cov}(X_k, X_j) = \operatorname{Cov}(Y, X_j), \quad j = 1, \ldots, n, \\ \displaystyle c_0^* = E[Y] - \sum_{k=1}^{n} c_k^* \, E[X_k]. \end{cases} \tag{6.2.2}$$

If the covariance matrix Λ of $X_1, \ldots, X_n$ is nonsingular, then the system can be solved either numerically or in a closed form through the inverse Λ^{-1} :

$$(c_1^*, \ldots, c_n^*)^\top = \Lambda^{-1} \left(\operatorname{Cov}(Y, X_1), \ldots, \operatorname{Cov}(Y, X_n)\right)^\top.$$

This is the main advantage of the least squares linear regression.

The system (6.2.2) shows that the problem (6.2.1) is solved in two steps: finding $c_1^*, \ldots, c_n^*$ and then determining c_0^*. In fact, for $\mathscr{D} = \sigma$, the error decomposition formulation (6.1.2) takes the form

$$(c_1^*, \ldots, c_n^*) \in \arg\min_{c_1,\ldots,c_n} \sigma\left(Y - \sum_{k=1}^{n} c_k X_k\right) \quad \text{and} \quad c_0 = E\left[Y - \sum_{k=1}^{n} c_k^* X_k\right],$$

which states that the least squares linear regression is equivalent to minimizing variance of $Y - \sum_{k=1}^{n} c_k X_k$ with respect to $c_1, \ldots, c_n$ and then setting intercept c_0 to the mean of optimal $Y - \sum_{k=1}^{n} c_k X_k$. This fact is often taken for granted and may create impression that the linear regression with another error measure $\mathscr{E}$ also leads to c_0 being $E[Y - \sum_{k=1}^{n} c_k^* X_k]$. However, this is possible only if the *deviation projected from* $\mathscr{E}$ coincides with the *deviation from the penalties relative to expectation*, i.e., if (2.4.3) coincides with (2.4.2); see Sect. 2.4.

The set of risk identifiers of X with respect to standard deviation is the singleton

$$\mathscr{Q}_\sigma(X) = \left\{1 - \frac{X - E[X]}{\sigma(X)}\right\},$$

and the optimality conditions (6.1.3) can be recast in the form

$$E\left[\left(Y - \sum_{k=1}^{n} c_k^* X_k\right)(X_j - E[X_j])\right] = 0, \qquad j = 1, \ldots, n,$$

which with $c_0^* = E\left[Y - \sum_{k=1}^{n} c_k^* X_k\right]$ are equivalent to the system (6.2.2).

In contrast to (6.0.2), the system (6.2.2) yields "true" $c_0, c_1, \ldots, c_n$ (not estimates) provided that the expected values $E[Y]$ and $E[X_k]$ and the covariances $\mathrm{Cov}(X_k, X_j)$ and $\mathrm{Cov}(Y, X_j)$ are known. However, in real-life problems, this is almost never the case: we are only given simultaneous observations of $X_1, \ldots, X_n$ and Y: $x_{1j}, \ldots, x_{nj}, y_j$, $j = 1, \ldots, m$, so that the expected values and covariances should be estimated through the given data.

6.2.2 Least Squares Linear Regression with Constraints

In applications, least squares linear regression is often solved subject to additional constraints on regression coefficients and, in general, can be formulated by

$$\min_{c_0,c_1,\ldots,c_n} \left\| Y - c_0 - \sum_{k=1}^{n} c_k X_k \right\|_2 \quad \text{subject to} \quad (c_0, c_1, \ldots, c_n) \in \mathscr{C}, \tag{6.2.3}$$

where $\mathscr{C}$ is some feasible set of $(c_0, c_1, \ldots, c_n)$. This problem admits a closed-form solution only in few simple cases, for example, when $\mathscr{C}$ is determined by a set of linear equalities. In a general case, (6.2.3) is solved numerically, for example, in Portfolio Safeguard with function st_pen.

Example 6.1 (index tracking with mean square error). Let Y be daily rate of return of a stock market index, e.g., S&P500 or Nasdaq, and let $X_1, \ldots, X_n$ be daily rates of return of chosen financial instruments. Suppose a unit capital is to be allocated among these instruments with capital weights $c_1, \ldots, c_n$ to replicate the index's rate of return by a linear combination of $X_1, \ldots, X_n$ without shorting of the instruments. The imposed requirements on $c_1, \ldots, c_n$ correspond to the feasible set

$$\mathscr{C} = \left\{ (c_1, \ldots, c_n) \in \mathbb{R}^n \;\middle|\; \sum_{k=1}^{n} c_k = 1, \;\; c_k \geq 0, \; k = 1, \ldots, n \right\}. \tag{6.2.4}$$

In this case, optimal allocation positions $c_1^*, \ldots, c_n^*$ can be found through the least squares linear regression (6.2.3) with $c_0 = 0$ and $\mathscr{C}$ given by (6.2.4), which is a quadratic optimization problem.

Example 6.2 (sparse reconstruction problem from SPARCO toolbox in "L2 D" formulation). Let $L\,(c, X)$ be an error function that linearly depends on a decision vector $c = (c_1, \ldots, c_n) \in \mathbb{R}^n$ and on a given random vector $X : \Omega \to \mathbb{R}^n$. The "L2 D" formulation of the sparse reconstruction problem from SPARCO toolbox is a regression that minimizes a linear combination of $\|L\,(c, X)\,\|_2^2$ and the regularization part $\sum_{i=1}^{n} |c_i|$ subject to box constraints $l_i \leq c_i \leq u_i$, $i = 1, \ldots, n$, where l_i and u_i are given bounds with $u_i \geq 0$ and $l_i \leq 0$. Let $c^{\pm} = (c_1^{\pm}, \ldots, c_n^{\pm})$ with $c_i^{\pm} = \max\{\pm c_i, 0\}$, then $c_i = c_i^{+} - c_i^{-}$ and $|c_i| = c_i^{+} + c_i^{-}$, and the "L2 D" formulation takes the form

$$\min_{c^{+}, c^{-}} \;\; \|L\,(c^{+} - c^{-}, X)\,\|_2^2 + \lambda \sum_{i=1}^{n} (c_i^{+} + c_i^{-})$$

$$\text{subject to } \; 0 \leq c_i^{+} \leq u_i, \quad 0 \leq c_i^{-} \leq -l_i, \quad i = 1, \ldots, n, \tag{6.2.5}$$

where λ is a given parameter.

In Sect. 9.12, the case study "Sparse Reconstruction Problems from SPARCO Toolbox" implements the problem (6.2.5) in Portfolio Safeguard and solves the problem with real-life data.

Example 6.3 (therapy treatment-planning problem). Let $[L(c, X)]_+ \equiv \max\{0, L(c, X)\}$ be a loss function, where $L(c, X)$ is an error function that linearly depends on a decision vector $c = (c_1, \ldots, c_n) \in \mathbb{R}^n$ and on a given random vector $X : \Omega \to \mathbb{R}^n$. The regression problem, arising in intensity-modulated radiation therapy treatment, minimizes $\|[L(c, X)]_+\|_2^2$ subject to box constraints $l_i \leq c_i \leq u_i$ with given bounds l_i and u_i:

$$\min_{c_1,\ldots,c_n} \quad \|[L(c,X)]_+\|_2^2 \quad \text{subject to} \quad l_i \le c_i \le u_i, \quad i = 1,\ldots,n. \tag{6.2.6}$$

In Sect. 9.13, the case study "Optimization of Intensity-Modulated Radiation Therapy Treatment-Planning Problem" implements the problem (6.2.6) in Portfolio Safeguard and solves the problem with real-life data.

6.3 Median Regression

In the least squares linear regression, large values of the error $Z = Y - c_0 - \sum_{k=1}^{n} c_k X_k$ are penalized heavier than small values, which makes the regression coefficients quite sensitive to outliers. In applications that require equal treatment of small and large errors, the median regression can be used instead.

6.3.1 Unconstrained Median Regression

Unconstrained median regression is a particular case of (6.1.1) with $\mathscr{E}(\cdot) = \|\cdot\|_1$:

$$\min_{c_0,c_1,\ldots,c_n} \left\| Y - c_0 - \sum_{k=1}^{n} c_k X_k \right\|_1, \tag{6.3.1}$$

for which optimal $c_0^*, c_1^*, \ldots, c_n^*$ are found by the error decomposition (6.1.2):

$$\begin{cases} (c_1^*,\ldots,c_n^*) \in \underset{c_1,\ldots,c_n}{\arg\min}\ E\left[|\widetilde{Z} - \operatorname{med}(\widetilde{Z})|\right], & \widetilde{Z} = Y - \sum_{k=1}^{n} c_k X_k, \\ c_0^* \in \operatorname{med}(\widetilde{Z}^*), & \widetilde{Z}^* = Y - \sum_{k=1}^{n} c_k^* X_k, \end{cases} \tag{6.3.2}$$

where $\operatorname{med}(\widetilde{Z})$ is the median of $\widetilde{Z}$, which, in general, is any number in the closed interval $[q_{\widetilde{Z}}^-(1/2), q_{\widetilde{Z}}^+(1/2)]$. Observe that the median regression does not reduce to minimization of the mean-absolute deviation (MAD) and that c_0^* is not the mean of $\widetilde{Z}^*$.

If the random variable $\widetilde{Z}^*$ has no probability "atom" at $q_{\widetilde{Z}^*}^+(1/2)$, then the interval $\operatorname{med}(\widetilde{Z}^*)$ is a singleton, and the optimality conditions (6.1.3) reduce to

$$E\left[X_j - E[X_j] \,\middle|\, \widetilde{Z}^* \le \operatorname{med}(\widetilde{Z}^*)\right] = 0, \quad j = 1,\ldots,n.$$

In applications, $X_1, \ldots, X_n$ and Y are often assumed to be discretely distributed with joint probability distribution $\mathbb{P}[X_1 = x_{1j}, \ldots, X_n = x_{nj}, Y = y_j] = p_j > 0$, $j = 1,\ldots,m$, with $\sum_{j=1}^{m} p_j = 1$. In this case,

$$\|Z\|_1 = \sum_{j=1}^{m} p_j \left| y_j - c_0 - \sum_{k=1}^{n} c_k x_{kj} \right| ,$$

and the median regression (6.3.1) reduces to the linear program

$$\begin{aligned} \min_{\substack{c_0, c_1, \ldots, c_n, \\ \zeta_1, \ldots, \zeta_m}} \quad & \sum_{j=1}^{m} p_j \zeta_j \\ \text{subject to} \quad & \zeta_j \geq y_j - c_0 - \sum_{k=1}^{n} c_k x_{kj}, \quad j = 1, \ldots, m, \\ & \zeta_j \geq c_0 + \sum_{k=1}^{n} c_k x_{kj} - y_j, \quad j = 1, \ldots, m, \end{aligned} \tag{6.3.3}$$

where $\zeta_1, \ldots, \zeta_m$ are auxiliary variables.

6.3.2 Median Regression with Constraints

The median regression with constraints on regression coefficients is formulated by

$$\min_{c_0, c_1, \ldots, c_n} \left\| Y - c_0 - \sum_{k=1}^{n} c_k X_k \right\|_1 \quad \text{subject to} \quad (c_0, c_1, \ldots, c_n) \in \mathscr{C}, \tag{6.3.4}$$

where $\mathscr{C}$ is a given feasible set of $(c_0, c_1, \ldots, c_n)$. For an arbitrary joint probability distribution of $X_1, \ldots, X_n$ and Y, the necessary optimality conditions for (6.3.4) are given in [42].

If $X_1, \ldots, X_n$ and Y are discretely distributed, and $\mathscr{C}$ is determined by a set of linear constraints, (6.3.4) reduces to a linear program. In this case, (6.3.4) can be solved in Portfolio Safeguard with the function meanabs_pen.

Example 6.4 (index tracking with mean absolute error). The problem setting is identical to that in Example 6.1. But this time, the optimal allocation positions $c_1^*, \ldots, c_n^*$ are found through the median regression (6.3.4) with $\mathscr{C}$ given by (6.2.4). If $X_1, \ldots, X_n$ and Y are assumed to be discretely distributed with joint probability distribution $\mathbb{P}[X_1 = x_{1j}, \ldots, X_n = x_{nj}, Y = y_j] = p_j > 0$, $j = 1, \ldots, m$, where $\sum_{j=1}^{m} p_j = 1$, then this regression problem can be formulated as the linear program

$$\begin{aligned} \min_{\substack{c_1,\ldots,c_n \\ \zeta_1,\ldots,\zeta_m}} \quad & \sum_{j=1}^{m} p_j \zeta_j \\ \text{subject to} \quad & \zeta_j \geq y_j - \sum_{k=1}^{n} c_k x_{kj}, \quad j = 1, \ldots, m, \\ & \zeta_j \geq \sum_{k=1}^{n} c_k x_{kj} - y_j, \quad j = 1, \ldots, m, \\ & \sum_{k=1}^{n} c_k = 1, \ c_k \geq 0, \quad k = 1, \ldots, n, \end{aligned}$$

where $\zeta_1, \ldots, \zeta_m$ are auxiliary variables.

Example 6.5 (median regression and collateralized debt obligation). Let $L(c, X)$ be a loss function in hedging against changes in a collateralized debt obligation (CDO) book, where $L(c, X)$ linearly depends on a decision vector $c = (c_1, \ldots, c_n) \in \mathbb{R}^n$ (positions in financial instruments) and on a given random vector $X : \Omega \to \mathbb{R}^n$. A regression problem then minimizes the mean absolute error of $L(c, X)$ subject to the budget constraint $\sum_{i=1}^{n} a_i |c_i| \leq C$ with given C and $a_i > 0$, $i = 1, \ldots, n$, and subject to a constraint on cardinality of the decision variables not to exceed a positive integer S:

$$\begin{aligned} \min_{c_1,\ldots,c_n} \quad & \|L(c, X)\|_1 \\ \text{subject to} \quad & \sum_{i=1}^{n} a_i |c_i| \leq C, \\ & \sum_{i=1}^{n} I_{\{a_i |c_i| \geq w\}} \leq S, \\ & |c_i| \leq k_i, \quad i = 1, \ldots, n, \end{aligned} \tag{6.3.5}$$

where $w \geq 0$ is a given threshold, $I_{\{\cdot\}}$ is the indicator function equal to 1 if the condition in the curly brackets is true and equal to 0 otherwise, and $|c_i| \leq k_i$, $i = 1, \ldots, n$, are bounds on decision variables (positions).

In Sect. 9.1, the case study "Optimal Hedging of CDO Book" implements the problem (6.3.5) in Portfolio Safeguard and solves the problem with real-life data.

Next three examples formulate regression problems arising in sparse signal reconstruction. In all of them, $L(c, X)$ is an error function that linearly depends on a decision vector $c = (c_1, \ldots, c_n) \in \mathbb{R}^n$ and on a given random vector $X : \Omega \to \mathbb{R}^n$, and $l_i \leq c_i \leq u_i$, $i = 1, \ldots, n$, are box constraints with given bounds l_i and u_i ($l_i \leq u_i$).

Example 6.6 (sparse signal reconstruction I: "L1Relaxed" formulation). This regression problem minimizes the mean absolute error of $L(c, X)$ subject to a constraint on cardinality of c with given bound $S \in \mathbb{N}$ and subject to box constraints on c:

$$\begin{aligned}
&\min_{c_1,\dots,c_n} \quad \|L(c, X)\|_1] \\
&\text{subject to} \quad \sum_{i=1}^{n} \left(I_{\{a_i c_i \geq w\}} + I_{\{b_i c_i \leq -w\}}\right) \leq S, \\
&\qquad\qquad\quad l_i \leq c_i \leq u_i, \quad i = 1, \dots, n.
\end{aligned} \tag{6.3.6}$$

Example 6.7 (sparse signal reconstruction II). This regression problem minimizes the mean absolute error of $L(c, X)$ subject to a constraint on the $\mathscr{L}_1$-norm of c, i.e., $\sum_{i=1}^{n} |c_i| \leq U$ with given bound $U \in \mathbb{R}^+$, and subject to box constraints on c:

$$\begin{aligned}
&\min_{c_1,\dots,c_n} \quad \|L(c, X)\|_1 \\
&\text{subject to} \quad \sum_{i=1}^{n} |c_i| \leq U, \\
&\qquad\qquad\quad l_i \leq c_i \leq u_i, \quad i = 1, \dots, n.
\end{aligned} \tag{6.3.7}$$

Example 6.8 (sparse signal reconstruction III). This estimation problem minimizes the cardinality of c subject to constraints on the mean absolute error of $L(c, X)$ and on the $\mathscr{L}_1$-norm of c with given bounds $\epsilon \in \mathbb{R}^+$ and $U \in \mathbb{R}^+$, respectively, and subject to box constraints on c:

$$\begin{aligned}
&\min_{c_1,\dots,c_n} \quad \sum_{i=1}^{n} \left(I_{\{a_i c_i \geq w\}} + I_{\{b_i c_i \leq -w\}}\right) \\
&\text{subject to} \quad \|L(c, X)\|_1 \leq \epsilon, \\
&\qquad\qquad\quad \sum_{i=1}^{n} |c_i| \leq U, \\
&\qquad\qquad\quad l_i \leq c_i \leq u_i, \quad i = 1, \dots, n.
\end{aligned} \tag{6.3.8}$$

In Sect. 9.11, the case study "Sparse Signal Reconstruction: a Cardinality Approach" implements the problems (6.3.6)–(6.3.8) in Portfolio Safeguard and solves them with real-life data.

The next example presents a reformulation of the regression problem (6.3.7).

Example 6.9 (sparse signal reconstruction from SPARCO toolbox in "L1Relaxed D" formulation). Suppose the random vector X is discretely distributed and takes on values $X^{(1)}, \dots, X^{(m)}$ with corresponding positive probabilities $p_1, \dots, p_m$ summing into 1, so that $\|L(c, X)\|_1 = \sum_{j=1}^{m} p_j \left|L(c, X^{(j)})\right|$. Let $c_i^{\pm} = \max\{\pm c_i, 0\}$,

$i = 1, \ldots, n$, then $c_i = c_i^+ - c_i^-$ and $|c_i| = c_i^+ + c_i^-$. Given that $L(c, X^{(j)})$ is linear with respect to c and that $u_i \geq 0$ and $l_i \leq 0$, $i = 1, \ldots, n$, the problem (6.3.7) can be restated as the linear program

$$\min_{\substack{c^+, c^- \\ \zeta_1, \ldots, \zeta_m}} \sum_{j=1}^{m} p_j \, \zeta_j$$

$$\begin{aligned}
\text{subject to } & \sum_{i=1}^{n} \left(c_i^+ + c_i^-\right) \leq U, \\
& \zeta_j \geq L\left(c^+ - c^-, X^{(j)}\right), \quad i = 1, \ldots, n, \\
& \zeta_j \geq -L\left(c^+ - c^-, X^{(j)}\right), \quad i = 1, \ldots, n, \\
& 0 \leq c_i^+ \leq u_i, \quad 0 \leq c_i^- \leq -l_i, \quad i = 1, \ldots, n,
\end{aligned} \tag{6.3.9}$$

where $\zeta_1, \ldots, \zeta_m$ are auxiliary variables and $c^\pm = (c_1^\pm, \ldots, c_n^\pm)$.

In Sect. 9.12, the case study "Sparse Reconstruction Problems from SPARCO Toolbox" implements the problem (6.3.9) in Portfolio Safeguard and solves the problem with real-life data.

6.4 Quantile Regression

Both the least squares linear regression and median regression treat ups and downs of the regression error equally, which might not be desirable in some applications. For example, in the index tracking problem from Example 6.1, a decision maker (financial analyst) may wish to penalize underperformance of the tracking portfolio heavier than its outperformance and, to this end, may use asymmetric error measures.

6.4.1 Unconstrained Quantile Regression

Unconstrained quantile regression is a particular case of the generalized linear regression (6.1.1) with the asymmetric mean absolute error measure (2.0.3):

$$\min_{c_0, c_1, \ldots, c_n} E\left[\alpha\, Z_+ + (1-\alpha) Z_-\right] \quad \text{with} \quad Z = Y - c_0 - \sum_{k=1}^{n} c_k X_k, \tag{6.4.1}$$

where $Z_\pm = \max\{\pm Z, 0\}$ and the multiplier α^{-1} in (2.0.3) is omitted. Observe that for $\alpha = 1/2$, (6.4.1) is equivalent to the median regression (6.3.1).

In this case, the error decomposition formulation (6.1.2) takes the form

$$(c_1^*,\ldots,c_n^*) \in \min_{c_1,\ldots,c_n} \mathrm{CVaR}_\alpha^\Delta\left(Y-\sum_{k=1}^n c_k X_k\right) \quad \text{and} \quad c_0^* \in \left[q_{\widetilde{Z}^*}^-(\alpha), q_{\widetilde{Z}^*}^+(\alpha)\right], \tag{6.4.2}$$

where $\widetilde{Z}^* = Y - \sum_{k=1}^n c_k^* X_k$. In other words, the quantile regression (6.4.1) reduces to minimizing *CVaR deviation* of $Y - \sum_{k=1}^n c_k X_k$ with respect to $c_1, \ldots, c_n$ and to setting c_0^* to any value from the α-quantile interval of $Y - \sum_{k=1}^n c_k^* X_k$.

If the random variable $\widetilde{Z}^*$ has no probability "atom" at $q_{\widetilde{Z}^*}^+(\alpha)$, then the interval $\left[q_{\widetilde{Z}^*}^-(\alpha), q_{\widetilde{Z}^*}^+(\alpha)\right]$ is a singleton, and the optimality conditions (6.1.3) simplify to

$$E\left[X_j - E[X_j] \,\middle|\, \widetilde{Z}^* \le q_{\widetilde{Z}^*}^+(\alpha)\right] = 0, \quad j = 1,\ldots,n.$$

However, in this form, they are rarely used in practice.

If $X_1, \ldots, X_n$ and Y are discretely distributed with joint probability distribution $\mathbb{P}[X_1 = x_{1j}, \ldots, X_n = x_{nj}, Y = y_j] = p_j > 0$, $j = 1, \ldots, m$, where $\sum_{j=1}^m p_j = 1$, then with the formula (1.4.4), the quantile regression (6.4.1) can be restated as the linear program

$$\min_{\substack{c_0,c_1,\ldots,c_n \\ \zeta_1,\ldots,\zeta_m}} \quad \sum_{j=1}^m p_j \left(y_j - c_0 + \alpha^{-1}\zeta_j - \sum_{k=1}^n c_k x_{kj}\right)$$

$$\text{subject to} \quad \zeta_j \ge c_0 + \sum_{k=1}^n c_k x_{kj} - y_j, \quad \zeta_j \ge 0, \quad j = 1,\ldots,m, \tag{6.4.3}$$

where $\zeta_1, \ldots, \zeta_m$ are auxiliary variables.

In Portfolio Safeguard, the quantile regression is implemented with pm_pen function, and there is no need to reduce the nonsmooth optimization problem (6.4.1) to (6.4.3). To solve (6.4.1) as a linear program, just specify "linearize = 1" in the code.

Next example presents an unconstrained quantile regression problem arising in the return-based style classification of a mutual fund.

Example 6.10 (quantile regression in style classification). Let $L(c, X)$ be a loss function that linearly depends on a decision vector $c \in \mathbb{R}^n$ and on a random vector $X : \Omega \to \mathbb{R}^n$ representing uncertain rates of return of n indices as explanatory variables. The quantile regression (6.4.1) with $L(c, X)$ in place of Z takes the form

$$\min_{c\in\mathbb{R}^n} E\left[\alpha[L(c,X)]_+ + (1-\alpha)[L(c,X)]_-\right]. \tag{6.4.4}$$

In Sect. 9.8, the case study "Style Classification with Quantile Regression" implements the quantile regression (6.4.4) in Portfolio Safeguard and solves it with real-life data.

6.4.2 Quantile Regression with Constraints

A constrained quantile regression is formulated similarly to (6.4.1):

$$\min_{c_0,c_1,\dots,c_n} \quad E[\alpha\, Z_+ + (1-\alpha)\, Z_-] \quad \text{with} \quad Z = Y - c_0 - \sum_{k=1}^{n} c_k X_k$$

$$\text{subject to} \quad (c_0, c_1, \dots, c_n) \in \mathscr{C}, \qquad (6.4.5)$$

where $\mathscr{C}$ is a given feasible set for regression coefficients $c_0, c_1, \dots, c_n$.

Example 6.11 (index tracking with asymmetric mean absolute error). The setting is identical to that in Example 6.1. But this time, the optimal allocation positions $c_1^*, \dots, c_n^*$ are found from the constrained quantile regression (6.4.5) with $\mathscr{C}$ given by (6.2.4). If $X_1, \dots, X_n$ and Y are assumed to be discretely distributed with joint probability distribution $\mathbb{P}[X_1 = x_{1j}, \dots, X_n = x_{nj}, Y = y_j] = p_j > 0$, $j = 1, \dots, m$, where $\sum_{j=1}^{m} p_j = 1$, then this regression problem can be formulated as the linear program

$$\min_{\substack{c_1,\dots,c_n \\ \zeta_1,\dots,\zeta_m}} \quad \sum_{j=1}^{m} p_j \left(y_j - \sum_{k=1}^{n} c_k x_{kj} + \alpha^{-1} \zeta_j \right)$$

$$\text{subject to} \quad \zeta_j \ge \sum_{k=1}^{n} c_k x_{kj} - y_j, \quad \zeta_j \ge 0, \quad j = 1, \dots, m,$$

$$\sum_{k=1}^{n} c_k = 1, \quad c_k \ge 0, \quad k = 1, \dots, n,$$

where $\zeta_1, \dots, \zeta_m$ are auxiliary variables.

6.5 Special Types of Linear Regression

This section discusses special types of unconstrained and constrained linear regressions encountered in statistical decision problems.

6.5.1 Mixed Quantile Regression

The linear regression with the *mixed quantile error measure* (2.4.5) in which positive weights $\lambda_1, \dots, \lambda_l$ such that $\sum_{k=1}^{l} \lambda_k = 1$ are given, is a generalization of quantile regression. For example, $\lambda_1, \dots, \lambda_l$ may collectively represent investor's risk profile corresponding to confidence levels $0 < \alpha_1 < \dots < \alpha_l < 1$.

Example 3.1 in [46] shows that the linear regression with (2.4.5) reduces to so-called mixed quantile regression

$$\min_{\substack{c_1,\ldots,c_n,\\ C_1,\ldots,C_l}} E\left[Y-\sum_{j=1}^{n} c_j X_j\right]+\sum_{k=1}^{l}\lambda_k\left(\frac{1}{\alpha_k}E\left[\max\left\{0, C_k-\sum_{j=1}^{n} c_j X_j\right\}\right]-C_k\right) \tag{6.5.1}$$

with the optimal intercept c_0^* determined by

$$c_0^* = \sum_{k=1}^{l} \lambda_k C_k^*,$$

where $C_1^*, \ldots, C_l^*$ are optimal in (6.5.1).

The optimality conditions (6.1.3) for (6.5.1) are complicated. However, as the quantile regression, (6.5.1) can be reduced to a linear program. Case Study 9.9 implements the mixed percentile regression (6.5.1) in Portfolio Safeguard with functions pm_pen and pm_pen_g.

6.5.2 *Unbiased Linear Regression*

Often, it is required to find an unbiased linear approximation of an output random variable Y by a linear combination of input random variables $X_1, \ldots, X_n$, in which case, the approximation error has zero expected value: $E[Y - c_0 - \sum_{k=1}^{n} c_k X_k] = 0$. A classical example of an *unbiased linear regression* is minimizing variance or, equivalently, standard deviation with the intercept c_0 set to the mean of optimal $Y - \sum_{k=1}^{n} c_k X_k$. If, in this example, the standard deviation is replaced by a general deviation measure $\mathscr{D}$, we obtain a generalized unbiased linear regression:

$$\min_{c_1,\ldots,c_n} \mathscr{D}\left(Y - \sum_{k=1}^{n} c_k X_k\right) \quad \text{and} \quad c_0^* = E\left[Y - \sum_{k=1}^{n} c_k^* X_k\right], \tag{6.5.2}$$

where $(c_1^*, \ldots, c_n^*) \in \arg\min_{c_1,\ldots,c_n} \mathscr{D}\left(Y - \sum_{k=1}^{n} c_k X_k\right)$.

In fact, (6.5.2) is equivalent to minimizing the error measure $\mathscr{E}(Z) = \mathscr{D}(Z) + |E[Z]|$ of $Z = Y - c_0 - \sum_{k=1}^{n} c_k X_k$. Observe that in view of the error decomposition theorem [46, Theorem 3.2], the generalized linear regression (6.1.1) with a nondegenerate error measure $\mathscr{E}$ and the unbiased linear regression (6.5.2) with the deviation measure $\mathscr{D}$ projected from $\mathscr{E}$ yield the same set of optimal $c_1^*, \ldots, c_n^*$ but, in general, different optimal intercepts c_0^*.

6.5.3 Risk Acceptable Regression

Rockafellar et al. [46] introduced *risk acceptable linear regression* in which a deviation measure $\mathcal{D}$ of the approximation error $Z = Y - c_0 - \sum_{k=1}^{n} c_k X_k$ is minimized subject to a constraint on the averse measure of risk $\mathcal{R}$ related to $\mathcal{D}$ by (2.3.1):

$$\min_{c_0,c_1,\dots,c_n} \quad \mathcal{D}(Z) \quad \text{subject to} \quad \mathcal{R}(Z) = 0 \quad \text{with} \quad Z = Y - c_0 - \sum_{k=1}^{n} c_k X_k, \tag{6.5.3}$$

which is equivalent to

$$\begin{cases} (c_1^*, \dots, c_n^*) \in \arg\min\limits_{c_1,\dots,c_n} \mathcal{D}\left(Y - \sum_{k=1}^{n} c_k X_k\right), \\ c_0^* = E\left[Y - \sum_{k=1}^{n} c_k^* X_k\right] - \mathcal{D}\left(Y - \sum_{k=1}^{n} c_k^* X_k\right). \end{cases} \tag{6.5.4}$$

The unbiased linear regression (6.5.2) and risk acceptable linear regression (6.5.4) show that the intercept c_0 could be set based on different requirements.

In general, the risk acceptable regression may minimize either an error measure $\mathcal{E}$ or a deviation measure $\mathcal{D}$ of the error Z subject to a constraint on a risk measure $\mathcal{R}$ of Z not necessarily related to $\mathcal{E}$ or $\mathcal{D}$. The next example illustrates a risk acceptable regression arising in a portfolio replication problem with a constraint on CVaR.

Example 6.12 (constrained risk acceptable regression). Let $L(c, X)$ be a portfolio replication error (loss function) that linearly depends on a decision vector $c \in \mathbb{R}^n$ and on a random vector $X : \Omega \to \mathbb{R}^n$ representing uncertain rates of return of n instruments in a portfolio replicating the S&P100 index. The constrained risk acceptable regression minimizes the mean absolute error of $L(c, X)$ subject to a CVaR constraint on the underperformance of the portfolio compared to the index, subject to the budget constraint $a^\top c \le U$ with given $a \in \mathbb{R}^n$ and $U \in \mathbb{R}^+$, and subject to no shorting constraints $c \ge 0$:

$$\begin{aligned} &\min_c \ \|L(c, X)\|_1 \\ &\text{subject to} \ \ \mathrm{CVaR}_\alpha(L(c, X)) \le w, \quad a^\top c \le U, \quad c \ge 0, \end{aligned} \tag{6.5.5}$$

where $\alpha \in [0, 1]$ and $w \in \mathbb{R}$ are given as well.

In Sect. 9.7, the case study "Portfolio Replication with Risk Constraint" implements the risk acceptable regression (6.5.5) in Portfolio Safeguard and solves it with real-life data.

6.5.4 Robust Regression

Robust regression aims to reduce influence of sample outliers on regression parameters, especially when regression error has heavy tails.

In statistics, robustness of an estimator is a well-established notion and is assessed by the so-called estimator's breakdown point, which is the proportion of additional arbitrarily large observations (outliers) needed to make the estimator unbounded. For example, the sample mean requires just a single such observation, while the sample median would still be finite until the proportion of such observations reaches 50 %. Consequently, mean's breakdown point is 0 %, whereas median's breakdown point is 50 %.

As in the previous regression setting, suppose Y is approximated by a linear combination of input random variables $X_1, \ldots, X_n$ with the regression error defined by $Z = Y - c_0 - \sum_{i=1}^{n} c_i X_i$, where $c_0, c_1, \ldots, c_n$ are unknown regression coefficients. Typically, a robust regression minimizes either an order statistic of some nonnegative function of Z or a "trimmed" error measure of Z that has nonzero breakdown point. In this case, regression's breakdown point is that of either the order statistic or the "trimmed" error measure.

Two popular robust regressions are *least median of squares (LMS) regression*, which minimizes the median of Z^2 and has 50 %-breakdown point:

$$\min_{c_0, c_1, \ldots, c_n} \operatorname{med}\left(Z^2\right) \quad \text{with} \quad Z = Y - c_0 - \sum_{i=1}^{n} c_i X_i, \tag{6.5.6}$$

and *least-trimmed squares (LTS) regression*, which minimizes the average α-quantile of Z^2 and has $(1-\alpha)\cdot 100$ %-breakdown point:

$$\min_{c_0, c_1, \ldots, c_n} \overline{q}_{Z^2}(\alpha) \quad \text{with} \quad Z = Y - c_0 - \sum_{i=1}^{n} c_i X_i. \tag{6.5.7}$$

Typically, in the LTS regression, α is set to be slightly larger than $1/2$. For $\alpha = 1$, $\overline{q}_{Z^2}(\alpha) = \|Z\|_2^2$, and (6.5.7) reduces to the standard least squares regression (i.e. $\overline{q}_{Z^2}(\alpha)$ is a "trimmed" least squares error measure). The LTS regression (6.5.7) is referred to as a challenging optimization problem [48] and is reported to have advantage over the LMS regression (6.5.6) or the one that minimizes the α-quantile of Z^2; see [48, 49, 59].

Let a function $h : \mathbb{R} \to \mathbb{R}_0^+$ be such that $h(t) > 0$ for $t \neq 0$ and $h(0) = 0$, but not necessarily symmetric, i.e., $h(-t) \neq h(t)$ in general. Then the LMS and LTS regressions have the following generalization:

(i) *Minimizing the upper α-quantile of $h(Z)$:*

$$\min_{c_0,c_1,\dots,c_n} q^+_{h(Z)}(\alpha) \quad \text{with} \quad Z = Y - c_0 - \sum_{i=1}^{n} c_i X_i. \tag{6.5.8}$$

(ii) *Minimizing the average α-quantile of $h(Z)$:*

$$\min_{c_0,c_1,\dots,c_n} \overline{q}_{h(Z)}(\alpha) \quad \text{with} \quad Z = Y - c_0 - \sum_{i=1}^{n} c_i X_i. \tag{6.5.9}$$

For example, in both (6.5.8) and (6.5.9), we may use $h(Z) = |Z|^p$, $p \geq 1$. In particular, for $h(Z) = Z^2$, (6.5.8) with $\alpha = 1/2$ corresponds to the LMS regression (6.5.6), whereas (6.5.9) reduces to the LTS regression (6.5.7).

When $h(-t) = h(t)$, (6.5.8) and (6.5.9) do not discriminate positive and negative errors. This, however, is unlikely to be appropriate for errors with significantly skewed distributions. For example, instead of med $\left(Z^2\right)$ and $\overline{q}_{Z^2}(\alpha)$, we can use *two-tailed α-value-at-risk (VaR) deviation* of the error Z defined by

$$\begin{aligned} \text{TwoTailVaR}^{\Delta}_{\alpha}(Z) &= \text{VaR}_{1-\alpha}(Z) + \text{VaR}_{1-\alpha}(-Z) \\ &\equiv q^-_Z(\alpha) - q^+_Z(1-\alpha), \qquad \alpha \in (1/2, 1]. \end{aligned} \tag{6.5.10}$$

The definition (6.5.10) shows that the two-tailed α-VaR deviation is, in fact, the range between the upper and lower $(1-\alpha)$-tails of the error Z, which is equivalent to the support of the random variable Z with truncated $(1-\alpha)\cdot 100\,\%$ of the "outperformances" and $(1-\alpha)\cdot 100\,\%$ of "underperformances." Consequently, the two-tailed α-VaR deviation has the breakdown point of $(1-\alpha)\cdot 100\,\%$. Typically, α is chosen to be 0.75 and 0.9. Next example formulates a robust regression with the two-tailed α-VaR deviation used in a mortgage pipeline hedging problem.

Example 6.13 (robust regression with two-tailed α-VaR deviation). Let a target random variable Y be approximated by a linear combination of n random variables $X_1, \dots, X_n$. In this case, the robust regression minimizes the two-tailed α-VaR deviation of the error $Y - c_0 - \sum_{i=1}^{n} c_i X_i$:

$$\min_{c_0,c_1,\dots,c_n} \text{TwoTailVaR}^{\Delta}_{\alpha}\left(Y - c_0 - \sum_{i=1}^{n} c_i X_i\right). \tag{6.5.11}$$

It has $(1-\alpha)\cdot 100\,\%$-breakdown point.

In Sect. 9.3, the case study "Mortgage Pipeline Hedging" formulates the problem (6.5.11) in Portfolio Safeguard and solves the problem with real-life data for $\alpha = 0.75$ and 0.9.

Chapter 7
Classification

This chapter discusses two classification methods: logistic regression and support vector machines (SVMs). Both methods are popular in various applications ranging from biomedicine and bioinformatics to image recognition and credit scoring. The logistic regression can classify a training data into several categories, whereas SVMs are mostly binary classifiers, i.e., deal with two classification categories.

7.1 Logistic Regression

Logistic regression aims to model a posterior discrete probability distribution of a random variable $Y\colon \Omega \to \mathbb{N}$ based on given independent observations of an m-dimensional random vector $X\colon \Omega \to \mathbb{R}^m$.

Suppose for any realization of X, the random variable Y may assume only K integer values $1, \ldots, K$, so that the posterior distribution of Y is given by

$$p_k(x) = \mathbb{P}[Y = k \mid X = x], \qquad k = 1, \ldots, K, \quad x \in \mathbb{R}^m.$$

One of the approaches to guarantee conditions $0 < p_k(x) < 1$ and $\sum_{k=1}^{K} p_k(x) = 1$ is to represent the probabilities $p_1(x), \ldots, p_K(x)$ in terms of *logits* $\phi_1(x), \ldots, \phi_{K-1}(x)$:

$$\begin{aligned} p_k(x) &= \frac{\exp(\phi_k(x))}{1 + \sum_{k=1}^{K-1} \exp(\phi_k(x))}, \qquad k = 1, \ldots, K-1, \\ p_K(x) &= \frac{1}{1 + \sum_{k=1}^{K-1} \exp(\phi_k(x))}. \end{aligned} \tag{7.1.1}$$

The inverse transformation of (7.1.1), also known as *logit transformation* or *log-odds*, is given by

M. Zabarankin and S. Uryasev, *Statistical Decision Problems: Selected Concepts and Portfolio Safeguard Case Studies*, Springer Optimization and Its Applications 85,
DOI 10.1007/978-1-4614-8471-4_7, © Springer Science+Business Media New York 2014

$$\phi_k(x) = \ln \frac{p_k(x)}{p_K(x)}, \qquad k = 1, \ldots, K-1. \tag{7.1.2}$$

Observe that in place of $p_K(x)$ in (7.1.1)–(7.1.2), we can use any of $p_1(x), \ldots, p_K(x)$.

The logistic regression approximates the logits $\phi_1(x), \ldots, \phi_{K-1}(x)$ *by linear functions of* x:

$$\phi_k(x) = c_{0k} + c_k^\top x, \qquad k = 1, \ldots, K-1,$$

where $c_{0k} \in \mathbb{R}$ and $c_k = (c_{1k}, \ldots, c_{mk}) \in \mathbb{R}^m$, $k = 1, \ldots, K-1$, are unknown coefficients.

Let $\theta = (c_{01}, c_{11}, \ldots, c_{m1}, \ldots, c_{0,K-1}, \ldots, c_{m,K-1})$, and let (x_i, y_i), $i = 1, \ldots, n$, be n independent observations of (X, Y) (training data), where $x_i \in \mathbb{R}^m$ and $y_i \in \{1, \ldots, K\}$. Then θ is found by maximizing the log-likelihood function for the observations $(x_1, y_1), \ldots, (x_n, y_n)$:

$$\max_\theta \sum_{i=1}^n \ln p_{y_i}(x_i; \theta),$$

where

$$\begin{aligned}
\ln p_{y_i}(x_i; \theta) &= \sum_{k=1}^{K-1} I_{\{k=y_i\}} \phi_k(x_i) - \ln\left(1 + \sum_{k=1}^{K-1} \exp(\phi_k(x_i))\right) \\
&= \sum_{k=1}^{K-1} I_{\{k=y_i\}} \left(c_{0k} + c_k^\top x_i\right) - \ln\left(1 + \sum_{k=1}^{K-1} \exp\left(c_{0k} + c_k^\top x_i\right)\right)
\end{aligned}$$

with $I_{\{\cdot\}}$ being the indicator function equal to 1 if the condition in the curly brackets is true and equal to 0 otherwise.

The logistic regression is often used in classification, where the values $1, \ldots, K$ represent different categories (groups). It arises in various applications, e.g., in bioinformatics and biomedicine that model probability/risk of certain events as a function of input random variables. The next section illustrates application of the logistic regression to classification for two categories (binary classification).

Logistic Regression for Two Categories

For $K = 2$, the logistic regression simplifies considerably. In this case, it is convenient to re-define the sample space of Y, which is $\{1, 2\}$, by $\{0, 1\}$, i.e., $1 \to 0$ and $2 \to 1$. Let $p_1(x) = p(x)$ and $\phi_1(x) = \phi(x)$; then the representation (7.1.1) reduces to

$$\mathbb{P}[Y = 0 \,|\, X = x] = p(x) = \frac{\exp(\phi(x))}{1 + \exp(\phi(x))},$$
$$\mathbb{P}[Y = 1 \,|\, X = x] = 1 - p(x) = \frac{1}{1 + \exp(\phi(x))}. \tag{7.1.3}$$

Observe that with Y assuming either 0 or 1, the logarithm of the distribution (7.1.3) can be represented in a concise form:

$$\begin{aligned} \ln \mathbb{P}[Y = y \,|\, X = x] &= (1 - y)\, \ln p(x) + y\, \ln(1 - p(x)) \\ &= (1 - y)\, \phi(x) - \ln\left(1 + \exp(\phi(x))\right). \end{aligned} \tag{7.1.4}$$

Now given n independent observations $(x_1, y_1), \ldots, (x_n, y_n)$ of (X, Y), where $x_i \in \mathbb{R}^m$ and $y_i \in \{0, 1\}$, $i = 1, \ldots, n$, and the linear approximation

$$\phi(x) = c_0 + c^\top x, \qquad c_0 \in \mathbb{R}, \quad c \in \mathbb{R}^m,$$

maximizing the log-likelihood function $\sum_{i=1}^n \ln p_{y_i}(x_i)$ with respect to c_0 and c takes the form

$$\max_{c_0, c} \sum_{i=1}^{n} \left((c_0 + c^\top x_i)(1 - y_i) - \ln\left(1 + \exp(c_0 + c^\top x_i)\right)\right). \tag{7.1.5}$$

In Portfolio Safeguard, the objective function in (7.1.5) is implemented by function logexp_sum, and therefore, solving (7.1.5) takes just few lines of code.

The necessary optimality conditions for (7.1.5) are given by the system

$$\begin{cases} \displaystyle\sum_{i=1}^{n} \left(y_i - \frac{1}{1 + \exp(c_0 + c^\top x_i)}\right) = 0, \\ \displaystyle\sum_{i=1}^{n} x_i \left(y_i - \frac{1}{1 + \exp(c_0 + c^\top x_i)}\right) = 0, \end{cases}$$

where the second equation is of dimension m, since $x_i \in \mathbb{R}^m$. This system can be solved with the *Newton–Raphson algorithm*, though, applying methods of numerical optimization to (7.1.5) directly could be more efficient.

Suppose $c_0^* \in \mathbb{R}$ and $c^* \in \mathbb{R}^m$ are optimal parameters in (7.1.5). Then for any realization $X = x \in \mathbb{R}^m$, the posterior distribution of Y is determined by (7.1.3) with $\phi(x) = c_0^* + c^{*\top} x$. If $1 - p(x) \geq p_0$, we decide that the corresponding (unknown) realization of Y is 1, otherwise the realization is 0. How to determine p_0? If we run the logistic regression for the original training data (used for finding optimal parameters c_0^* and c^*) and apply this classification rule, then the number of observations classified to be in category 2 should coincide with the true number of the observations in category 2, which is $\sum_{i=1}^n y_i$. Consequently, p_0 is determined by

$$p_0 = \sup \left\{ t \in [0,1] \,\middle|\, \sum_{i=1}^{n} I_{\{1-p(x_i)\geq t\}} \geq \sum_{i=1}^{n} y_i \right\}$$

or, equivalently,

$$p_0 = \sup \left\{ t \in [0,1] \,\middle|\, \sum_{i=1}^{n} I_{\{c_0^* + c^{*\top} x_i \leq \ln \frac{1-t}{t}\}} \geq \sum_{i=1}^{n} y_i \right\}. \tag{7.1.6}$$

Then for a new observation x of X, the classification rule for the corresponding realization y of Y is given by

$$\text{If} \quad \phi(x) = c_0^* + c^{*\top} x > \ln[(1-p_0)/p_0], \quad \text{then } y = 0, \quad \text{otherwise } y = 1. \tag{7.1.7}$$

Example 7.1 (Logistic regression in disease diagnostics). Let (x_i, y_i), $i = 1, \ldots, n$, represent medical data of n patients susceptible to a certain disease, where $x_i \in \mathbb{R}^m$ is a vector of test results and y_i is 0 if patient i is healthy and 1 if patient i is sick. Now, we fit the logistic regression for $K = 2$ with the data, i.e., solve the problem (7.1.5), and use the classification rule (7.1.7) to deduce whether a new patient with given test results x is either healthy or sick.

The case study in Sect. 9.18 implements the problem (7.1.5) with Portfolio Safeguard and solves the problem with real-life data.

7.2 Support Vector Machine

SVM is a binary classification technique, which maps given training data from two categories into a multidimensional space and then separates the sets of points representing the two categories by a hyperplane with maximal margin. If a mapping is linear, the SVM is called *linear*, otherwise it is called *nonlinear*. In the case of linear SVM, if a separating hyperplane exists, the training data is said to be linearly separable. This section reviews typical formulations of the SVM.

7.2.1 *Standard SVM*

Suppose training data from two categories (1 and 2) are given in the form (x_i, y_i), $i = 1, \ldots, n$, where $x_i \in \mathbb{R}^m$ and $y_i = 1$ if x_i belongs to category 1 and $y_i = -1$ if x_i belongs to category 2. Let $\phi : \mathbb{R}^m \to \mathbb{R}^l$ be a mapping applied to $x_1, \ldots, x_n$, so that the transformed training data are given by $(\phi(x_i), y_i)$, $i = 1, \ldots, n$ (typically, $l \geq m$). The SVM objective is to separate points $\phi(x_i)$ with $y_i = 1$ from points $\phi(x_i)$ with $y_i = -1$ by a hyperplane with maximal margin, namely,

$$\max_{w,\, b_1,\, b_2} \frac{|b_1 - b_2|}{\|w\|}$$

$$\text{subject to} \quad \phi(x_i) \cdot w + b_1 \geq 0 \quad \text{if} \quad y_i = 1,$$

$$\phi(x_i) \cdot w + b_2 \leq 0 \quad \text{if} \quad y_i = -1,$$

where w is an l-dimensional vector perpendicular to the hyperplane, $\|w\|$ is the length of w, and "·" is the dot (scalar) product. Observe that optimal w, b_1, and b_2 are determined up to the same positive multiplier. This "degree of freedom" can be eliminated by setting $b_1 = b - 1$ and $b_2 = b + 1$, so that $|b_1 - b_2|/\|w\| = 2/\|w\|$ and the SVM problem is reformulated in the standard concise form

$$\min_{w,\, b} \frac{1}{2}\|w\|^2 \quad \text{subject to} \quad y_i(\phi(x_i) \cdot w + b) \geq 1, \quad i = 1, \ldots, n, \tag{7.2.1}$$

which is a quadratic optimization problem. Points $\phi(x_i)$ for which the constraint in (7.2.1) is active (holds as equality) are called *support vectors*, and $1/\|w\|$ is called *separation margin* (distance from the support vectors to the separating hyperplane).

Suppose w^* and b^* solve (7.2.1), then $f(x) = \phi(x) \cdot w^* + b^*$ is the decision (classification) function for a new observation $x \in \mathbb{R}^m$: the category of x is determined by $y = \operatorname{sign}(f(x))$.

7.2.2 SVM with Soft Margin

If (7.2.1) has no solution, the training data is not separable with chosen ϕ. In this case, the SVM can be reformulated as *C-SVM* with *soft margin*, which allows violating the constraint in (7.2.1):

$$\min_{w,\, b,\, \tau_1,\ldots,\tau_n} \left(\frac{1}{2}\|w\|^2 + C\sum_{i=1}^{n} \tau_i\right)$$

$$\text{subject to} \quad y_i(\phi(x_i) \cdot w + b) \geq 1 - \tau_i, \quad \tau_i \geq 0, \quad i = 1, \ldots, n, \tag{7.2.2}$$

where a nonnegative parameter C controls the extent to which the constraint violation is penalized; see [8]. The C-SVM trades off the reciprocal of the square margin and the separation error. In this case, (7.2.1) is also referred to as the SVM with *hard margin*.

The problem (7.2.2) can be recast in an unconstrained form

$$\min_{w,\, b} \left(\frac{1}{2}\|w\|^2 + C\sum_{i=1}^{n} [1 - y_i(\phi(x_i) \cdot w + b)]_+\right), \tag{7.2.3}$$

where the second term is interpreted as the expected value of the loss function $[t]_+ \equiv \max\{t, 0\}$ with multiplier $n\,C$. This interpretation suggests that other downside measures can be used in place of $[t]_+$.

One of the well-known extensions of the C-SVM is ν-*SVM* formulated by

$$\min_{w,\,b,\,\rho\geq 0} \left(\frac{1}{2}\|w\|^2 - \nu\rho + \frac{1}{n}\sum_{i=1}^{n} [\rho - y_i(\phi(x_i)\cdot w + b)]_+ \right) \tag{7.2.4}$$

with $\nu \in [0, 1]$ being an upper bound for the fraction of constraint violations. Crisp and Burges [12] showed that the constraint $\rho \geq 0$ is redundant, and Schölkopf et al. [52] proved that if optimal ρ is positive, then (7.2.4) is equivalent to (7.2.3) with $C = 1/(n\rho)$.

With the constraint $\rho \geq 0$ in (7.2.4) being dropped, the formulae (1.4.4) and (2.0.2) imply that

$$\min_{\rho} \left(-\rho + \frac{1}{n\,\nu}\sum_{i=1}^{n} [\rho - y_i(\phi(x_i)\cdot w + b)]_+ \right) = -\overline{q}_Z(\nu) \equiv \mathrm{CVaR}_\nu(Z), \tag{7.2.5}$$

where $Z = y(\phi(x)\cdot w + b)$ is interpreted as a discretely distributed random variable assuming values $y_1(\phi(x_1)\cdot w + b), \dots, y_n(\phi(x_n)\cdot w + b)$ with equal probabilities $1/n$.

Consequently, the ν-SVM (7.2.4) can be recast in the form

$$\min_{w,\,b} \left(\frac{1}{2}\|w\|^2 + \nu\,\mathrm{CVaR}_\nu\,(y(\phi(x)\cdot w + b)) \right). \tag{7.2.6}$$

Chang and Lin [5] proved that there exist $\nu_{\min}$ and $\nu_{\max}$ ($0 \leq \nu_{\min} \leq \nu_{\max} \leq 1$) such that

- If $\nu \in [0, \nu_{\min}]$, then (7.2.4) has only trivial solution ($w = 0$, $b = 0$).
- If $\nu \in (\nu_{\min}, \nu_{\max}]$, then (7.2.4) is equivalent to (7.2.3) with some C.
- If $\nu \in (\nu_{\max}, 1]$, then (7.2.4) has no solution.

Also, Crisp and Burges [12] showed that $\nu_{\max} = 2\min\{k_1, k_2\}$, where k_1 and k_2 are the fractions of the training data in the two categories.

In Sect. 9.16, the case study "ν-SVM with Conditional Value-at-Risk" implements the problem (7.2.6) in Portfolio Safeguard and solves the problem with real-life data.

To address the issue of trivial solution of (7.2.4) for $\nu \in [0, \nu_{\min}]$, Perez-Cruz et al. [35] modified the ν-SVM (7.2.4) by dropping $\rho \geq 0$ and imposing the constraint on w to be a unit vector:

$$\min_{w,\,b,\,\rho}\ \left(-\nu\rho + \frac{1}{n}\sum_{i=1}^{n}[\rho - y_i(\phi(x_i)\cdot w + b)]_+\right) \quad \text{subject to} \quad \|w\| = 1. \tag{7.2.7}$$

This problem is known as *extended* ν*-SVM* (Eν-SVM), and with (7.2.5), it can be rewritten in the form

$$\min_{w,\,b}\ \mathrm{CVaR}_\nu\,(y(\phi(x)\cdot w + b)) \quad \text{subject to} \quad \|w\| = 1. \tag{7.2.8}$$

The Eν-SVM is reported to outperform the ν-SVM (see [35]). Also, the Eν-SVM has a clear interpretation: it minimizes the average of $\nu \cdot 100\,\%$ of the worst classification errors. However, the drawback of the Eν-SVM is that it is nonconvex because of the constraint $\|w\| = 1$.

Let w^* and b^* be an optimal solution of (7.2.8) for given ν, and let $\overline{\nu}$ be such that $\mathrm{CVaR}_{\overline{\nu}}\,(y(\phi(x)\cdot w^* + b^*)) = 0$. Takeda and Sugiyama [55, Lemmas 6 and 9] showed that for $\nu \in (\overline{\nu}, \nu_{\max}]$, the equality constraint $\|w\| = 1$ in (7.2.8) can be replaced by the inequality $\|w\| \le 1$, and (7.2.8) becomes convex, whereas for $\nu \in (0, \overline{\nu}]$, $\|w\| = 1$ can be replaced by $\|w\| \ge 1$, and (7.2.8) becomes a *linear reverse convex program* (LRCP).

Another extension of the ν-SVM (7.2.6), called *robust* ν*-SVM*, assumes that there are M training datasets $\left\{\left(x_j^{(k)}, y_j^{(k)}\right)\right\}_{j=1}^{n_k}$, $k = 1,\dots,M$, where $x_j^{(k)} \in \mathbb{R}^m$ and $y_j^{(k)} \in \{0,1\}$ for $j = 1,\dots,n_k$ and $k = 1,\dots,M$. It optimizes the separation margin for the worst-case scenario (dataset):

$$\min_{w,\,b}\left(\frac{1}{2}\|w\|^2 + \nu \max_{k=1,\dots,M} \mathrm{CVaR}_\nu\left(y^{(k)}\left(\phi\left(x^{(k)}\right)\cdot w + b\right)\right)\right), \tag{7.2.9}$$

where $y^{(k)}\left(\phi\left(x^{(k)}\right)\cdot w + b\right)$ is interpreted as a discretely distributed random variable assuming values $y_1^{(k)}\left(\phi\left(x_1^{(k)}\right)\cdot w + b\right),\dots, y_{n_k}^{(k)}\left(\phi\left(x_{n_k}^{(k)}\right)\cdot w + b\right)$ with equal probabilities $1/n_k$.

In Sect. 9.16, the case study "ν-SVM with Conditional Value-at-Risk" implements the problem (7.2.9) in Portfolio Safeguard and solves the problem with real-life data.

7.2.3 *Alternative Formulations of SVM with Soft Margin*

Given training data $(x_1, y_1),\dots,(x_n, y_n)$, where $x_i \in \mathbb{R}^m$ and $y_i \in \{-1, 1\}$ for $i = 1,\dots,n$, let $y(\phi(x)\cdot w + b)$ be interpreted as a discretely distributed random variable assuming values $y_1(\phi(x_1)\cdot w + b),\dots, y_n(\phi(x_n)\cdot w + b)$ with equal probabilities $1/n$. With this interpretation, the "hard-margin" SVM (7.2.1) can be equivalently rewritten as

$$\min_{w,\,b} \ \frac{1}{2}\|w\|^2 \quad \text{subject to} \quad \mathbb{P}\left[y(\phi(x)\cdot w + b) - 1 \geq 0\right] = 1.$$

Now if we allow the constraint $y(\phi(x)\cdot w + b) - 1 \geq 0$ to be violated for $\alpha \cdot 100\,\%$ of the training data, where $\alpha = [0, 1)$, then the SVM with "soft margin" can be formulated by

$$\min_{w,\,b} \ \frac{1}{2}\|w\|^2 \quad \text{subject to} \quad \mathbb{P}\left[y(\phi(x)\cdot w + b) - 1 \leq 0\right] \leq \alpha,$$

which implies

$$\min_{w,\,b} \ \frac{1}{2}\|w\|^2 \quad \text{subject to} \quad \mathrm{VaR}_\alpha\left(y(\phi(x)\cdot w + b) - 1\right) \leq 0, \tag{7.2.10}$$

where VaR_α is the negative upper α-quantile defined by (2.0.1). VaR-SVM (7.2.10) is shown to be stable to data outliers [58], which makes it attractive for robust classification. However, (7.2.10) has two deficiencies, which both stem from using the α-quantile. Namely, it is nonconvex, and while it guarantees that the constraint $y(\phi(x)\cdot w + b) - 1 \geq 0$ is violated for no more than $\alpha \cdot 100\,\%$ of the training data, where $\alpha = [0, 1)$, the extent of the violations can be quite significant.

Another approach to formulate an SVM with "soft margin" is to rewrite (7.2.1) in the form

$$\min_{w,\,b} \ \frac{1}{2}\|w\|^2 \quad \text{subject to} \quad \mathrm{CVaR}_0\left(y(\phi(x)\cdot w + b) - 1\right) \leq 0, \tag{7.2.11}$$

where CVaR_α is the negative average α-quantile defined by (2.0.2) and $\mathrm{CVaR}_0(y(\phi(x)\cdot w + b) - 1) = -\min_{1\leq i\leq n}(y_i(\phi(x_i)\cdot w + b) - 1)$. If (7.2.11) has no solution, we can change CVaR_0 to CVaR_α with some $\alpha \in [0, 1)$, so that yet another SVM with "soft margin" is given by

$$\min_{w,\,b} \ \frac{1}{2}\|w\|^2 \quad \text{subject to} \quad \mathrm{CVaR}_\alpha\left(y(\phi(x)\cdot w + b) - 1\right) \leq 0, \tag{7.2.12}$$

which is a convex optimization problem. There is no guarantee, though, that for chosen α, (7.2.12) has a solution. However, it has a solution (w^*, b^*) for $\alpha = 1$: if $\sum_{i=1}^n y_i \neq 0$, then $w^* = 0$ and b^* is such that $b^* \sum_{i=1}^n y_i \geq n$; and if $\sum_{i=1}^n y_i = 0$, then $w^* = n \sum_{i=1}^n y_i\phi(x_i) \Big/ \left\|\sum_{i=1}^n y_i\phi(x_i)\right\|^2$ and b^* is arbitrary. Consequently, there exists $\alpha^* \leq 1$ such that for $\alpha \in [\alpha^*, 1]$, the problem (7.2.12) has a solution.

Observe that the constraint in (7.2.12) is equivalent to $\mathrm{CVaR}_\alpha(y(\phi(x)\cdot w + b)) \leq -1$, and consequently, (7.2.12) with $\alpha \in [\alpha^*, 1]$ is closely related to the Eν-SVM (7.2.8) with $\nu \in (\overline{\nu}, \nu_{\max}]$.

7.2.4 Choice of Mapping ϕ: Kernel Approach

Now the question is: what ϕ should be used? Often ϕ is chosen to be a polynomial function. In particular, for $\phi(x) = x$, (7.2.1) reduces to the classical linear SVM. However, the analysis of the dual of (7.2.1) shows that explicit knowledge of ϕ is, in fact, not necessary. Indeed, the dual implies that if $\lambda_i^* \in \mathbb{R}_0^+$ is the optimal Lagrange multiplier corresponding to the ith constraint in (7.2.1), then optimal w can be represented by

$$w^* = \sum_{j=1}^{n} \lambda_j^* \, y_j \, \phi(x_j), \qquad (7.2.13)$$

and the decision function $f(x)$ takes the form

$$f(x) = \sum_{j=1}^{n} \lambda_j^* \, y_j \, \phi(x_j) \cdot \phi(x) + b^*,$$

where b^* is found from any constraint in (7.2.1) for which $\lambda_i^* \neq 0$. Consequently, we merely need to know how to compute the dot product in the transformed space.

A generalization of Mercer's theorem [31] implies that for two vectors $x \in \mathbb{R}^m$ and $x' \in \mathbb{R}^m$, the dot product of $\phi(x)$ and $\phi(x')$ can be evaluated through some *non-negative definite kernel* K:[1]

$$K(x, x') = \phi(x) \cdot \phi(x').$$

There are three widely used kernels:

Polynomial kernel	$K(x, x') = (x \cdot x' + c)^d$	$c \geq 0, \;\; d \in \mathbb{N}$
Neural network kernel	$K(x, x') = \tanh(c_1 \, x \cdot x' + c_2)$	$c_1 > 0, \;\; c_2 < 0$
Radial basis function (RBF) kernel	$K(x, x') = \exp\left(-\gamma \, \|x - x'\|^2\right)$	$\gamma > 0$

The RBF kernel is, arguably, the most popular choice.

In terms of kernel K, the decision function $f(x)$ of the SVM (7.2.1) takes the form

$$f(x) = \sum_{j=1}^{n} \lambda_j^* \, y_j \, K(x, x_j) + b^*, \qquad (7.2.14)$$

[1]Kernel $K = K(x, x')$ is a symmetric continuous real-valued function of $x \in \mathbb{R}^m$ and $x' \in \mathbb{R}^m$, and it is non-negative definite if $\sum_{i=1}^n \sum_{j=1}^n a_i \, a_j \, K(x_i, x_j) \geq 0$ for any $x_i \in \mathbb{R}^m$ and any $a_i \in \mathbb{R}$, $i = 1, \ldots, n$.

where $\lambda_1^*, \ldots, \lambda_n^*$ are an optimal solution of the dual of (7.2.1):

$$\begin{aligned} \max_{\lambda_1,\ldots,\lambda_n} \quad & \sum_{i=1}^n \lambda_i - \frac{1}{2} \sum_{i=1}^n \sum_{j=1}^n \lambda_i \lambda_j y_i y_j K(x_i, x_j) \\ \text{subject to} \quad & \sum_{i=1}^n \lambda_i y_i = 0, \quad \lambda_i \geq 0, \quad i = 1, \ldots, n, \end{aligned} \tag{7.2.15}$$

and the complementary slackness conditions $\lambda_i^*(y_i(\phi(x_i) \cdot w^* + b^*) - 1) = 0$, $i = 1, \ldots, n$, imply that $b^* = y_{i_1} - \sum_{j=1}^n \lambda_j^* y_j K(x_{i_1}, x_j)$ if there is $i_1 \in \{1, \ldots, n\}$ such that $\lambda_{i_1}^* \neq 0$.

For the C-SVM (7.2.2), optimal w is deteremined by (7.2.13), and the decision function $f(x)$ coincides with (7.2.14), where this time, the optimal Lagrange multipliers $\lambda_1^*, \ldots, \lambda_n^*$ correspond to the constraints $y_i(\phi(x_i) \cdot w + b) \geq 1 - \tau_i$, $i = 1, \ldots, n$, and solve the dual of the C-SVM (7.2.2):

$$\begin{aligned} \max_{\lambda_1,\ldots,\lambda_n} \quad & \sum_{i=1}^n \lambda_i - \frac{1}{2} \sum_{i=1}^n \sum_{j=1}^n \lambda_i \lambda_j y_i y_j K(x_i, x_j) \\ \text{subject to} \quad & \sum_{i=1}^n \lambda_i y_i = 0, \quad 0 \leq \lambda_i \leq C, \quad i = 1, \ldots, n. \end{aligned} \tag{7.2.16}$$

In this case, the complementary slackness conditions $\lambda_i^*(C - \lambda_i^*)(y_i(\phi(x_i) \cdot w^* + b^*) - 1) = 0$, $i = 1, \ldots, n$, yield $b^* = y_{i_1} - \sum_{j=1}^n \lambda_j^* y_j K(x_{i_1}, x_j)$ provided that there is $i_1 \in \{1, \ldots, n\}$ such that $\lambda_{i_1}^* \neq 0$ and $\lambda_{i_1}^* \neq C$.

For the ν-SVM (7.2.4), the kernel approach works as follows. First, (7.2.4) is reformulated by

$$\begin{aligned} \min_{w,\, b,\, \rho,\, \tau_1,\ldots,\tau_n} \quad & \left(\frac{1}{2}\|w\|^2 - \nu\rho + \frac{1}{n} \sum_{i=1}^n \tau_i \right) \\ \text{subject to} \quad & \tau_i \geq \rho - y_i(\phi(x_i) \cdot w + b), \quad \tau_i \geq 0, \quad i = 1, \ldots, n, \end{aligned} \tag{7.2.17}$$

where $\tau_1, \ldots, \tau_n$ are auxiliary variables and the constraint $\rho \geq 0$ was dropped in view of its redundancy (see [12]). With Lagrange multiplier $\lambda_i \in \mathbb{R}_0^+$ corresponding to the constraint $\tau_i \geq \rho - y_i(\phi(x_i) \cdot w + b)$, $i = 1, \ldots, n$, and with $K(x_i, x_j) = \phi(x_i) \cdot \phi(x_j)$, the dual of (7.2.17) takes the form

$$\max_{\lambda_1,\dots,\lambda_n} \quad -\frac{1}{2}\sum_{i=1}^{n}\sum_{j=1}^{n}\lambda_i\,\lambda_j\,y_i\,y_j\,K(x_i,x_j)$$

$$\text{subject to} \quad \sum_{i=1}^{n}\lambda_i = \nu, \quad \sum_{i=1}^{n}\lambda_i\,y_i = 0, \tag{7.2.18}$$

$$0 \le \lambda_i \le \frac{1}{n}, \quad i = 1,\dots,n.$$

Let $\lambda_1^*,\dots,\lambda_n^*$ be an optimal solution of (7.2.18), then optimal w in (7.2.17) is determined by (7.2.13), and the decision function $f(x)$ coincides with (7.2.14), where b^* is found from the complementary slackness conditions $\lambda_i^*(\frac{1}{n}-\lambda_i^*)(\rho^* - y_i(\phi(x_i)\cdot w^* + b^*)) = 0, i = 1,\dots,n$, with ρ^* being optimal ρ in (7.2.17). Suppose there are $i_1, i_2 \in \{1,\dots,n\}$ such that $y_{i_1} \ne y_{i_2}$ and $\lambda_i^*\left(\frac{1}{n}-\lambda_i^*\right) \ne 0$ for $i = i_1, i_2$, then $b^* = -\frac{1}{2}\sum_{j=1}^{n}\lambda_j^*\,y_j\,(K(x_{i_1},x_j) + K(x_{i_2},x_j))$.

Chapter 8
Statistical Decision Models with Risk and Deviation

Statistical decision problems with deviation and risk measures naturally arise in various financial engineering applications such as portfolio selection, risk hedging, index tracking, and factor analysis. For example, Markowitz's portfolio selection problem minimizes variance, or, equivalently, standard deviation, of the portfolio rate of return subject to a constraint on the expected value of the portfolio rate of return.

Let X be a random variable that depends on a decision vector $x \in \mathbb{R}^n$ and on a given random vector θ, i.e., $X = X(x, \theta)$, and let $\mathscr{X}$ be a feasible set of X determined by direct constraints on x, e.g., by $x \geq 0$. In general, a statistical decision problem can involve several risk and deviation measures of X in any combination with each other and with other functionals, e.g., $E[X]$ and $\|X\|_p$, in both the objective function and constraints subject to $X \in \mathscr{X}$. This chapter demonstrates a variety of problem formulations with risk and deviation measures encountered in real-life applications.

8.1 Minimization of Risk and Deviation

There are two basic problems with risk and deviation measures that deserve special consideration, namely, minimization of a deviation measure $\mathscr{D}$

$$\min_{X \in \mathscr{X}} \mathscr{D}(X) \tag{8.1.1}$$

and minimization of an *averse measure of risk* $\mathscr{R}$

$$\min_{X \in \mathscr{X}} \mathscr{R}(X) \tag{8.1.2}$$

with $\mathscr{D}$ and $\mathscr{R}$ defined in Sects. 2.1 and 2.3, respectively, and with the feasible set $\mathscr{X}$ in both problems given by

M. Zabarankin and S. Uryasev, *Statistical Decision Problems: Selected Concepts and Portfolio Safeguard Case Studies*, Springer Optimization and Its Applications 85, DOI 10.1007/978-1-4614-8471-4_8, © Springer Science+Business Media New York 2014

$$\mathscr{X} = \{X = x^\top \theta |\ Ax \le b\}, \tag{8.1.3}$$

where $x \in \mathbb{R}^n$ is a decision vector, $\theta : \Omega \to \mathbb{R}^n$ is a given random vector on a sample space Ω, and $b \in \mathbb{R}^m$ and A and are given constant vector and $m \times n$ matrix with constant entries $a_{ij} \in \mathbb{R}$, $i = 1, \ldots, n$, $j = 1, \ldots, m$, respectively. The optimality conditions for (8.1.1) and (8.1.2) are elaborated in [41], whereas here, we briefly summarize main results.

First, we introduce key notions of convex analysis: *convex set*, *convex functional*, *subdifferential*, and *normal cone*; see, e.g., [36].

Definition 8.1 (Convex set). A set $\mathscr{X}$ is convex if $\lambda X_1 + (1 - \lambda)X_2 \in \mathscr{X}$ for any two elements $X_1 \in \mathscr{X}$ and $X_2 \in \mathscr{X}$ and for any $\lambda \in [0, 1]$.

Definition 8.2 (Closed set). In a topological space, a set $\mathscr{X}$ is closed if and only if it contains all its limit points.

Definition 8.3 (Convex functional). A functional $\mathscr{F} : \mathscr{X} \to \mathbb{R}$ is convex on a convex set $\mathscr{X}$ if $\mathscr{F}(\lambda X_1 + (1 - \lambda)X_2) \le \lambda \mathscr{F}(X_1) + (1 - \lambda)\mathscr{F}(X_2)$ for any two elements $X_1 \in \mathscr{X}$ and $X_2 \in \mathscr{X}$ and for any $\lambda \in [0, 1]$.

Definition 8.4 (Subdifferential). The *subdifferential* for an arbitrary convex functional $\mathscr{F} : \mathscr{L}^2 \to (-\infty, \infty]$ at a point X^* is defined by

$$\partial \mathscr{F}(X^*) = \{Y \in \mathscr{L}^2 \,|\, \mathscr{F}(X) \ge \mathscr{F}(X^*) + E[(X - X^*)Y] \text{ for all } X \in \mathscr{L}^2\}.$$

The subdifferential is a closed and convex set. It reduces to the singleton consisting of the gradient $\nabla \mathscr{F}(X^*)$ when $\mathscr{F}$ is differentiable at X^* (in the Fréchet sense) and is empty when $\mathscr{F}(X^*) = \infty$.

Definition 8.5 (Normal cone). For an arbitrary nonempty, closed and convex set $\mathscr{X} \subseteq \mathscr{L}^2$, the *normal cone* at a point $X^* \in \mathscr{X}$ is defined by

$$N_{\mathscr{X}}(X^*) = \{Y \in \mathscr{L}^2 |\ E[(X - X^*)Y] \le 0 \text{ for all } X \in \mathscr{X}\}.$$

The normal cone is always closed and convex and contains 0.

For $\mathscr{X}$ given by (8.1.3), the normal cone $N_{\mathscr{X}}(X^*)$ is characterized as follows. For any element X^* of $\mathscr{X}$, corresponding to $x^* \in \mathbb{R}^n$ in (8.1.3) such that $Ax^* \le b$, $Y \in N_{\mathscr{X}}(X^*)$ if and only if there exist a multiplier vector $\lambda \in \mathbb{R}^m$ satisfying

$$E[\theta\, Y] = \lambda^\top A, \quad \lambda \ge 0, \quad \lambda^\top (Ax^* - b) = 0.$$

The following theorem is restated from [41, Theorem 5].

Theorem 8.1 (Optimality conditions for risk or deviation). *Let $\mathscr{D}$, $\mathscr{R}$, and $\mathscr{Q}$ be a deviation measure, the averse measure of risk related to $\mathscr{D}$ by* (2.3.1), *and the risk envelope associated with $\mathscr{D}$, respectively*[1] *(see Sect. 2.2). For each $X \in \mathscr{L}^2$, let*

[1] In [41, Theorem 5], $\mathscr{D}$ and $\mathscr{R}$ are required to be lower semicontinuous, whereas here, they are lower semicontinuous by definition.

$$\mathcal{Q}_{\mathcal{D}}(X) = \arg\min_{Q \in \mathcal{Q}} E[XQ] = \big\{Q \in \mathcal{Q} \,\big|\, E[XQ'] \geq E[XQ] \;\; \textit{for all} \;\; Q' \in \mathcal{Q}\big\}$$

be the set of risk identifiers associated with $\mathcal{D}$ (see Sect. 2.2). The subdifferentials for $\mathcal{D}$ and $\mathcal{R}$ then take the form

$$\partial\mathcal{R}(X) = \{-Q \,|\, Q \in \mathcal{Q}_{\mathcal{D}}(X)\} = \{-Q \,|\, Q \in \mathcal{Q},\; \mathcal{R}(X) = -E[XQ]\},$$

$$\partial\mathcal{D}(X) = \{Y \,|\, 1 - Y \in \mathcal{Q}_{\mathcal{D}}(X)\} = \{Y \,|\, 1 - Y \in \mathcal{Q},\; \mathcal{D}(X) = E[XY]\}.$$

In particular, $Y \in \partial\mathcal{D}(X)$ if and only if $Y - 1 \in \partial\mathcal{R}(X)$.

If $\mathcal{X}$ is nonempty, closed, and convex, then

(a) A sufficient condition for X^* to be an optimal solution to (8.1.1)*, which is also necessary when $\mathcal{D}$ is finite everywhere on $\mathcal{L}^2$, is the existence of*

$$Y^* \in \partial\mathcal{D}(X^*) \quad \textit{such that} \quad -Y^* \in N_{\mathcal{X}}(X^*).$$

(b) A sufficient condition for X^* to be an optimal solution to (8.1.2)*, which is also necessary when $\mathcal{R}$ is finite everywhere on $\mathcal{L}^2$, is the existence of*

$$-Q^* \in \partial\mathcal{R}(X^*) \quad \textit{such that} \quad Q^* \in N_{\mathcal{X}}(X^*).$$

The averse measure of risk $\mathcal{R}$ and deviation measure $\mathcal{D}$ related by (2.3.1) are often used interchangeably. For example, conditional value-at-risk (CVaR) is often minimized in place of CVaR-deviation. Theorem 4 in [41] elucidates when optimal solutions of (8.1.1) and (8.1.2) with such $\mathcal{D}$ and $\mathcal{R}$ coincide. It relies on the notion of *recession cone* [36].

Definition 8.6 (Recession cone). The *recession cone* $\mathrm{rc}\mathcal{X}$ of an arbitrary nonempty, closed, and convex set $\mathcal{X} \subseteq \mathcal{L}^2$ is given by

$$\mathrm{rc}\mathcal{X} = \big\{X' \in \mathcal{L}^2 \,\big|\, X + X' \in \mathcal{X} \quad \text{for any} \quad X \in \mathcal{X}\big\}.$$

It is closed and convex (see [36]).

For example, for $\mathcal{X}$ defined by (8.1.3), $\mathrm{rc}\mathcal{X} = \{X = x^{\top}\theta \,|\, Ax \leq 0\}$.

Theorem 8.2 (Risk minimization versus deviation minimization).

(a) If the elements $X \in \mathcal{X}$ have to satisfy an equality constraint $E[X] = \mu$ for some given μ, the problems (8.1.1) *and* (8.1.2) *are equivalent in the sense that they have the same set of optimal solutions and optimal values are related by*

$$\mathcal{D}(X^*) = \mathcal{R}(X^*) + \mu$$

for any optimal X^*.

(b) *If the elements $X \in \mathscr{X}$ only have to satisfy an inequality constraint $E[X] \geq \mu$ for some given μ, the problems* (8.1.1) *and* (8.1.2) *are equivalent in the same sense as long as it can be established that their optimal solutions must actually satisfy this inequality constraint with equality (i.e., this constraint must be* active *in optimality). Otherwise, these problems can differ. In particular, if there exists $X' \in \mathrm{rc}\mathscr{X}$ such that $\mathscr{R}(X') < 0$, then the problem* (8.1.2) *has no solution with $\mathscr{R}(X)$ being unbounded from below, whereas the problem* (8.1.1) *may still have a solution.*

8.2 Deviation Minimization in Portfolio Selection

8.2.1 Markowitz Portfolio Selection

Markowitz portfolio selection [29, 30] is one of the most well-known decision models in finance, in which standard deviation of the portfolio rate of return or, equivalently, variance is minimized subject to constraints on portfolio weights (decision vector). It is a particular case of (8.1.1) and is formulated as follows. Suppose there is a risk-free instrument with the rate of return r_0, and suppose there are n risky instruments with rates of return given by a random vector $r = (r_1, \ldots, r_n)$ having covariance matrix Λ. Let x_i be the portion of a unit capital invested into instrument i, $i = 0, 1, \ldots, n$, and let $x = (x_1, \ldots, x_n)$ be the vector of portfolio weights. Then the portfolio rate of return is determined by $X = \sum_{i=0}^{n} r_i x_i \equiv r_0 x_0 + r^\top x$, and the problem is to minimize variance $\sigma^2(X) = x^\top \Lambda x$ subject to the budget constraint $\sum_{i=0}^{n} x_i = 1$ and a constraint on the expected value of X not to drop below given value $\mu > r_0$ or formally

$$\begin{aligned} &\min_{x_0, x} \; x^\top \Lambda x \\ &\text{subject to} \; E[r_0 x_0 + r^\top x] \geq \mu, \\ &\qquad\qquad x_0 + e^\top x = 1, \end{aligned} \tag{8.2.1}$$

where $e = (1, \ldots, 1)$ is the n-dimensional unit vector.

The optimality conditions for the problem (8.2.1) reduce to a system of linear equations, which has a closed-form solution provided that Λ has the inverse. Indeed, it can be shown that the constraint $E[r_0 x_0 + r^\top x] \geq \mu$ is always active, and with the Lagrange multipliers technique, we obtain optimal values of x and x_0 as functions of μ:

$$\begin{aligned} \widetilde{x}(\mu) &= \frac{(\mu - r_0)\Lambda^{-1}(E[r] - r_0 e)}{(E[r] - r_0 e)^\top \Lambda^{-1}(E[r] - r_0 e)}, \\ \widetilde{x}_0(\mu) &= 1 - e^\top \widetilde{x}(\mu). \end{aligned} \tag{8.2.2}$$

If μ is such that the optimal portfolio has zero weight in the risk-free asset, i.e., $\widetilde{x}_0(\mu) = 0$, then the portfolio is called master fund of positive type, which is also known as *market portfolio*. The condition $\widetilde{x}_0(\mu) = 0$ along with (8.2.2) implies that the market portfolio has the expected rate of return

$$\mu_M = r_0 + \frac{(E[r] - r_0 e)^\top \Lambda^{-1} (E[r] - r_0 e)}{e^\top \Lambda^{-1} (E[r] - r_0 e)}, \tag{8.2.3}$$

and, consequently, the weights and variance of the market portfolio are determined by

$$x^* = \frac{\Lambda^{-1} (E[r] - r_0 e)}{e^\top \Lambda^{-1} (E[r] - r_0 e)}, \qquad x_0^* = 0, \tag{8.2.4}$$

$$\sigma_M^2 = \frac{(E[r] - r_0 e)^\top \Lambda^{-1} (E[r] - r_0 e)}{(e^\top \Lambda^{-1} (E[r] - r_0 e))^2}. \tag{8.2.5}$$

The formulae (8.2.3)–(8.2.5) have two important implications.

Corollary 8.1 (One-fund theorem). *With* (8.2.3)–(8.2.5)*, the solution* (8.2.2) *can be represented in the form:*

$$\widetilde{x}(\mu) = \frac{\mu - r_0}{\mu_M - r_0} x^*, \qquad \widetilde{x}_0(\mu) = 1 - \frac{\mu - r_0}{\mu_M - r_0}, \tag{8.2.6}$$

which is the well-known one-fund theorem [53, 57]. In other words, an optimal investment strategy is to invest the fraction $(\mu - r_0)/(\mu_M - r_0)$ of the initial capital into the market portfolio and the rest into the risk-free asset. If $\mu > \mu_M$, then $\widetilde{x}_0(\mu) = (\mu_M - \mu)/(\mu_M - r_0) < 0$, in which case, an investor should borrow at the risk-free rate and invest the whole capital into the market portfolio.

Corollary 8.2 (Capital asset pricing model (CAPM)). *It follows from* (8.2.3) *and* (8.2.5) *that*

$$e^\top \Lambda^{-1} (E[r] - r_0 e) = \frac{\mu_M - r_0}{\sigma_M^2}$$

and the first formula in (8.2.4) *can be rearranged in the form*

$$E[r] - r_0 e = \frac{\mu_M - r_0}{\sigma_M^2} \Lambda x^*.$$

Let $r_M = r^\top x^*$ be the rate of return of the market portfolio. Then $\Lambda x^* = (\mathrm{Cov}(r_1, r_M), \ldots, \mathrm{Cov}(r_n, r_M))^\top$, and the last expression can be rewritten as

$$E[r_i] - r_0 = \beta_i (E[r_M] - r_0), \qquad \beta_i = \frac{\mathrm{Cov}(r_i, r_M)}{\sigma_M^2}, \qquad i = 1, \ldots, n, \tag{8.2.7}$$

which is the well-known CAPM; see [14, 53, 54].

Suppose now the rate of return of the market portfolio is used as an *explanatory variable* for the rates of return of the risky instruments in a *one-factor asset pricing model*:

$$r_i - r_0 = a_i (r_M - r_0) + b_i + \epsilon_i, \qquad i = 1, \ldots, n, \tag{8.2.8}$$

where ϵ_i is an error and a_i and b_i are parameters found from the least squares linear regression:

$$a_i = \beta_i, \qquad b_i = E[r_i] - r_0 - \beta_i \left(E[r_M] - r_0 \right), \qquad i = 1, \ldots, n,$$

where β_i is the same as in (8.2.7). This shows that the CAPM (8.2.7) corresponds to the intercepts $b_1, \ldots, b_n$ set to zero in the one-factor linear model (8.2.8).

If shorting is not allowed, the Markowitz portfolio selection problem may no longer have a closed-form solution. In this case, solving (8.2.1) subject to $x_0 \geq 0$ and $x \geq 0$ is resorted to numerical optimization techniques.

Example 8.1 (Markowitz portfolio selection with no shorting and no risk-free asset). If in the introduced setting, shorting is not allowed and the risk-free asset is not available, the Markowitz portfolio selection problem takes the form

$$\min_x \; x^\top \Lambda x \quad \text{subject to} \quad E[r^\top x] \geq \mu, \quad e^\top x = 1, \quad x \geq 0. \tag{8.2.9}$$

In Sect. 9.14, the case study "Portfolio Optimization: CVaR versus Standard Deviation" implements the problem (8.2.9) in Portfolio Safeguard in two cases: when the covariance matrix Λ is available and when the rates of return of the instruments are modeled by scenarios.

8.2.2 Markowitz-Type Portfolio Selection

Let the portfolio setting be identical to that in Sect. 8.2.1, and let the feasible set for the portfolio rate of return X be given by

$$\mathscr{X} = \left\{ X = x_0 r_0 + r^\top x \;\middle|\; x_0 + e^\top x = 1 \right\}.$$

A Markowitz-type portfolio selection [43–45] is the problem (8.2.1), in which the variance of X is replaced by an arbitrary deviation measure $\mathscr{D}$, and is formulated by

$$\min_{X \in \mathscr{X}} \; \mathscr{D}(X) \quad \text{subject to} \quad E[X] \geq r_0 + \Delta, \tag{8.2.10}$$

where $\Delta > 0$ is the desired gain over the risk-free rate r_0.

The problem (8.2.10) is considered under the following assumptions:

(A1) No portfolio with $x \neq 0$ is risk free.
(A2) The expected rates of return $E[r_1], \ldots, E[r_n]$ are not all the same.
(A3) $\mathscr{D}(r_i) < \infty$ and $\mathscr{D}(-r_i) < \infty$ for $i = 1, \ldots, n$.

Assumption A1 is satisfied if and only if none of the instruments can be replicated through the others and holds when $r_1, \ldots, r_n$ are multivariate continuously distributed [43, Proposition 2], whereas Assumption A3 along with convexity of $\mathscr{D}$ guarantees that $\mathscr{D}(X) < \infty$ for all bounded $X \in \mathscr{X}$. Assumption A2 excludes the degenerate case when any portfolio of only risky instruments has the same expected rate of return. Its value will be evident from the discussion of the role of the risk-free rate of return r_0.

Expressing x_0 from the budget constraint

$$x_0 = 1 - e^\top x$$

and rescaling x by Δ, i.e., $x := x/\Delta$, we can restate (8.2.10) in the form

$$\min_{x \in \mathbb{R}^n} \mathscr{D}\left(r^\top x\right) \quad \text{subject to} \quad x^\top \left(E[r] - r_0\, e\right) \geq 1. \tag{8.2.11}$$

The sufficient condition for x^* to be optimal in (8.2.11) is given by Theorem 4 in [44]:

$$\operatorname{Cov}\left(-r_i, Q^*\right) = \mathscr{D}(r^\top x^*)\left(E[r_i] - r_0\right), \qquad i = 1, \ldots, n, \tag{8.2.12}$$

provided that there exists $Q^* \in \mathscr{Q}_{\mathscr{D}}(r^\top x^*)$, where $\mathscr{Q}_{\mathscr{D}}(r^\top x^*)$ is the set of risk identifiers of $r^\top x^*$ with respect to deviation $\mathscr{D}$. If $\mathscr{D}$ is finite and continuous,[2] then (8.2.12) is also the necessary optimality condition.

Let x^* be an optimal solution of (8.2.11). Then the optimal portfolio weights in (8.2.10) are functions of Δ determined by

$$\widetilde{x}(\Delta) = \Delta\, x^*, \qquad \widetilde{x}_0(\Delta) = 1 - \Delta\, e^\top x^*. \tag{8.2.13}$$

Now, what is a *market portfolio* in this case, and can the optimal investment policy determined by (8.2.12)–(8.2.13) be formulated in the form similar to the one-fund theorem (8.2.6)? If $\Delta = 1/(e^\top x^*) > 0$, then the optimal portfolio has the weights $\widetilde{x}_0(\Delta) = 0$ and $\widetilde{x}(\Delta) = x^*/(e^\top x^*)$ and is called *master fund of positive type* or *market portfolio*. However, Theorem 5 in [43] shows that there exists a threshold interval $[r_0^-, r_0^+]$ for the risk-free rate of return r_0 such that $e^\top x^* > 0$ if $r_0 < r_0^-$, $e^\top x^* = 0$ if $r_0^- < r_0 < r_0^+$, and $e^\top x^* < 0$ if $r_0 > r_0^+$. In other words,

[2]Since a deviation measure is lower semicontinuous by property D4, and since a finite lower semicontinuous deviation measure is continuous [42, Proposition 2], we only need to require finiteness of $\mathscr{D}$, i.e., that $\mathscr{D}(X) < \infty$ for any $X \in \mathscr{L}^2(\Omega)$.

if r_0 is set too high (greater than r_0^-), then no master fund of positive type exists, in which case, the notion of "market portfolio" loses its meaning. The threshold values r_0^- and r_0^+ are determined by

$$r_0^- = -\frac{\rho^+}{d*}, \qquad r_0^+ = -\frac{\rho^-}{d*},$$

where ρ^- and ρ^+ are the minimal and maximal values of the optimal Lagrange multiplier ρ in the Lagrangian $L(x, \rho, \eta) = \mathscr{D}(r^\top x) + \rho[-e^\top x] + \eta[1 - E[r^\top x]]$ of the auxiliary problem

$$d^* = \min_x \mathscr{D}(r^\top x) \quad \text{subject to} \quad e^\top x = 0, \quad E[r^\top x] = 1,$$

which is guaranteed to have a solution by Assumption A2 (for $n \geq 2$), and d^* is the optimal value of this problem; see [43] for details. For example, for the standard deviation,

$$r_0^- = r_0^+ = \frac{e^\top \Lambda^{-1} E[r]}{e^\top \Lambda^{-1} e},$$

provided that Λ^{-1} exists (Λ is the covariance matrix of $r_1, \ldots, r_n$). Consequently, the classical one-fund theorem (8.2.6) and classical CAPM (8.2.7) hold only if $r_0 < \left(e^\top \Lambda^{-1} E[r]\right) / \left(e^\top \Lambda^{-1} e\right)$.

If $e^\top x^* < 0$, the portfolio with the weights $x_0 = 0$ and $x = x^* /(-e^\top x^*)$ is called *master fund of negative type* (its price is -1), whereas if $e^\top x^* = 0$, the portfolio with the weights $x_0 = 0$ and $x = x^*$ is called *master fund of threshold type* (its price is 0).

It follows from Theorems 3 and 6 in [43] that

- When $r_0 < r_0^-$, there is a master fund of positive type with the rate of return $r_M = r^\top x^* /(e^\top x^*)$ but none of negative type, so invest the amount $\Delta/(E[r_M] - r_0)$ in the master fund and the amount $1 - \Delta/(E[r_M] - r_0)$ in the risk-free instrument.
- When $r_0 \in (r_0^-, r_0^+)$, there is a master fund of a threshold type with the rate of return $r_M = r^\top x^*$, but there is neither a master fund of positive type nor one of negative type, so invest the whole capital in the risk-free instrument and open a position of magnitude Δ in the master fund through long and short positions.
- When $r_0 > r_0^+$, there is a master fund of negative type with the rate of return $r_M = r^\top x^* /(-e^\top x^*)$, but there is none of positive type, so invest $-\Delta/(E[r_M] - r_0)$ in the master fund and $1 + \Delta/(E[r_M] - r_0)$ in the risk-free instrument.
- When $r = r_0^-$ or $r = r_0^+$, there exists a master fund of threshold type, but when $r = r_0^-$, there could also exist a master fund of positive type, whereas when $r = r_0^+$, there could also exist a master fund of negative type, so that the optimal investment policy is formulated similarly to the previous three cases.

This result generalizes the classical one-fund theorem (8.2.6) stated for the case of the standard deviation when a master fund of positive type (market portfolio) exists.

Special Cases

If $r_1, \ldots, r_n$ are *jointly normally distributed* with the vector of means $\mu \in \mathbb{R}^n$ and covariance matrix Λ, then the portfolio rate of return $X = r_0 x_0 + r^\top x$ is normally distributed with mean $\mu(X) = r_0 x_0 + \mu^\top x$ and variance $\sigma^2(X) = x^\top \Lambda x$, and for any deviation measure, we have

$$\mathscr{D}(X) = \sigma(X)\, \mathscr{D}(Z),$$

where Z is a standard normal random variable. In this case, the Markowitz-type portfolio problem (8.2.10) simplifies to (8.2.1) and, thus, has the same solution regardless of the chosen deviation measure. Also, VaR deviation $\mathrm{VaR}_\alpha^\Delta(X) = E[X] + \mathrm{VaR}_\alpha(X)$, which is not, in fact, a deviation measure (it lacks convexity), takes the form $\mathrm{VaR}_\alpha^\Delta(X) = -\sigma(X)\mathrm{VaR}_\alpha(Z)$, and (8.2.10) with $\mathscr{D} = \mathrm{VaR}_\alpha^\Delta$ and $\alpha \in (0, 1/2)$ reduces to (8.2.1) as well.

If shorting is not allowed and the risk-free asset is not available, then the portfolio rate of return is defined by $X = r^\top x$ with the budget constraint $e^\top x = 1$, and the Markowitz-type portfolio selection problem with an arbitrary deviation measure $\mathscr{D}$ takes the form

$$\begin{aligned} &\min_{X \in \mathscr{X}} \ \mathscr{D}(X) \quad \text{subject to} \quad E[X] \geq r_0 + \Delta, \\ &\text{with} \quad \mathscr{X} = \left\{ X = r^\top x \ \middle|\ e^\top x = 1, \ \ x \geq 0 \right\}. \end{aligned} \tag{8.2.14}$$

In Sect. 9.14, the case study "Portfolio Optimization: CVaR versus Standard Deviation" implements the problem (8.2.14) in Portfolio Safeguard for CVaR deviation and solves the problem with real-life data.

8.2.3 Generalized Capital Asset Pricing Model

The optimality condition (8.2.12) for the Markowitz-type portfolio problem (8.2.10) has the following important interpretation. If $r_0 < r_0^-$, then there exists a master fund of positive type with the weights $x_0 = 0$ and $x = x^* / (e^\top x^*)$ and with the rate of return $r_M = r^\top x^* / (e^\top x^*)$, where x^* is a solution to (8.2.12). Then, on the one hand, (8.2.10) implies that $E[r_M] - r_0 = \Delta$, and, on the other hand, $\Delta = 1 / (e^\top x^*)$ by the definition of a master fund of positive type, so that $e^\top x^* = 1/(E[r_M] - r_0)$ and $r^\top x^* = r_M\, e^\top x^* = r_M / (E[r_M] - r_0)$, and (8.2.12) can be recast in the form

$$E[r_i] - r_0 = \frac{\mathrm{Cov}\,(-r_i, Q^*)}{\mathscr{D}(r_M)}\,(E[r_M] - r_0)\,, \qquad i = 1, \ldots, n.$$

For an arbitrary case of r_0, Theorem 5 in [44] shows that the optimality condition (8.2.12) can be formulated as *CAPM-like relations* that generalize the classical CAPM (8.2.7):

$$E[r_i] - r_0 \\ = \begin{cases} \beta_i (E[r_M] - r_0), & \text{when there exists a master fund of positive type,} \\ \beta_i (E[r_M] + r_0), & \text{when there exists a master fund of negative type,} \\ \beta_i E[r_M], & \text{when there exists a master fund of threshold type,} \end{cases} \tag{8.2.15}$$

where r_M is the rate of return of a master fund and β_i is the *generalized beta* of instrument i determined by

$$\beta_i = \frac{\mathrm{Cov}(-r_i, Q_M)}{\mathscr{D}(r_M)}, \qquad Q_M \in \mathscr{Q}_{\mathscr{D}}(r_M), \quad i = 1, \ldots, n. \tag{8.2.16}$$

Example 8.2 (Standard deviation). If $\mathscr{D}(X) = \sigma(X)$, then

$$\mathscr{Q}_\sigma(r_M) = \left\{1 - \frac{r_M - E[r_M]}{\sigma(r_M)}\right\}$$

and

$$\beta_i = \frac{\mathrm{Cov}(r_i, r_M)}{\sigma^2(r_M)},$$

which is the classical beta in (8.2.7).

Example 8.3 (Standard lower semideviation). If $\mathscr{D}(X) = \sigma_-(X)$, then

$$\mathscr{Q}_{\sigma_-}(X) = \left\{1 - \frac{E[[E[r_M] - r_M]_+] - [E[r_M] - r_M]_+}{\sigma_-(r_M)}\right\}$$

and

$$\beta_i = \frac{\mathrm{Cov}(-r_i, [E[r_M] - r_M]_+)}{\sigma_-^2(r_M)}.$$

Example 8.4 (CVaR deviation). If $\mathscr{D}(X) = \mathrm{CVaR}_\alpha^\Delta(X)$, then $\mathscr{Q}_{\mathrm{CVaR}_\alpha^\Delta}(r_M)$ is determined by (2.2.3) with $X = r_M$, and

$$\beta_i = \frac{E[(E[r_i] - r_i)Q_M]}{\mathrm{CVaR}_\alpha^\Delta(r_M)}, \qquad Q_M \in \mathscr{Q}_{\mathrm{CVaR}_\alpha^\Delta}(r_M). \tag{8.2.17}$$

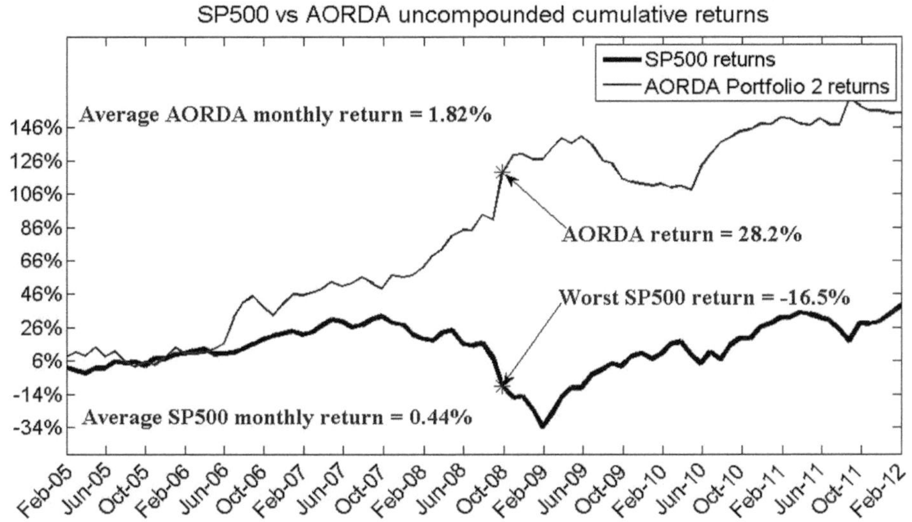

Fig. 8.1 Uncompounded monthly cumulative rates of return of the S&P500 index and AORDA portfolio for February 2005–February 2012: *asterisk* indicates when S&P500's lowest rate of return occurs

When $\mathbb{P}[r_M = -\text{VaR}_\alpha(r_M)] = 0$, (8.2.17), can be expressed in terms of conditional probabilities

$$\beta_i = \frac{E[E[r_i] - r_i \mid r_M \leq -\text{VaR}_\alpha(r_M)]}{E[E[r_M] - r_M \mid r_M \leq -\text{VaR}_\alpha(r_M)]}.$$

The formula (8.2.17) for β_i can be illustrated with the following example. Suppose the S&P500 index and an AORDA portfolio represent the master fund and instrument i in (8.2.17), respectively, and suppose the monthly rates of return of these two instruments are given for the period February 2005–February 2012 (85 months) with the average values of 0.44 % and 1.82 %, respectively; see Fig. 8.1. Then for $\text{CVaR}_\alpha^\Delta$ with $\alpha = 0$, which corresponds to the lower range deviation, $\text{CVaR}_0^\Delta(r_M) = 0.44\,\% - (-16.5\,\%)$, where $-16.5\,\%$ is the lowest S&P500's rate of return for February 2005–February 2012 (occurred in October 2008; see Fig. 8.1) and $E[(E[r_i] - r_i)Q_M] = 1.82\,\% - 28.2\,\%$, where 28.2 % is the AORDA portfolio rate of return that corresponds to the S&P500's rate of return of $-16.5\,\%$ (October 2008). Consequently,

$$\beta_i = \frac{1.82\,\% - 28.2\,\%}{0.44\,\% - (-16.5\,\%)} = -1.56.$$

As another illustration for the formula (8.2.17), suppose in the same example $\alpha = 0.977$, so that $(1 - 0.977) \times 85 \approx 2$. In this case, $\text{CVaR}_{0.977}^\Delta(r_M) = 0.44\,\% - 0.5(-16.5\,\% - 10.75\,\%)$, where $0.5(-16.5\,\% - 10.75\,\%)$ is the average

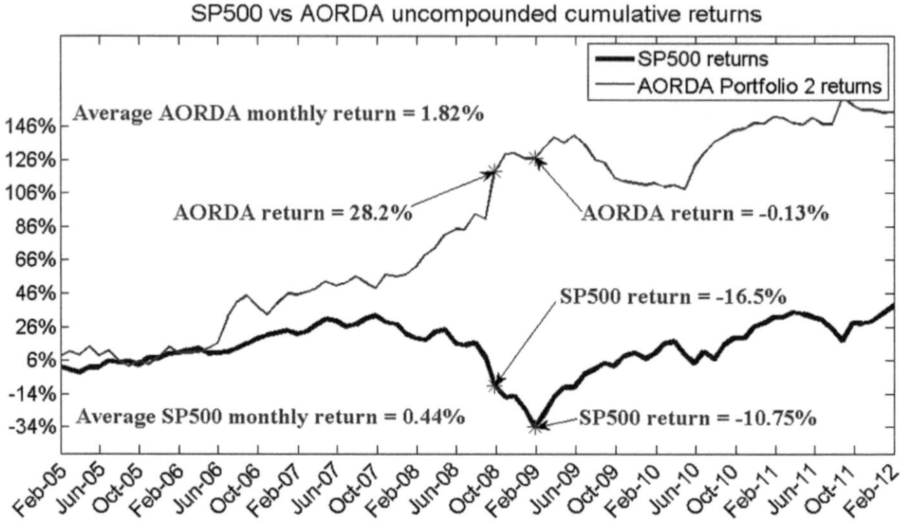

Fig. 8.2 Uncompounded monthly cumulative rates of return of the S&P500 index and AORDA portfolio for February 2005–February 2012: *asterisks* indicate when two S&P500's lowest rates of return occur

of the two lowest S&P500's rates of return for the period February 2005–February 2012 (occurred in October 2008 and February 2009; see Fig. 8.2), and $E[(E[r_i] - r_i)Q_M] = 1.82\,\% - 0.5(28.2\,\% - 0.13\,\%)$, where the AORDA portfolio rates of return of 28.2 % and −0.13 % correspond to the two lowest S&P500's rates of return (October 2008 and February 2009). Consequently,

$$\beta_i = \frac{1.82\,\% - 0.5(28.2\,\% - 0.13\,\%)}{0.44\,\% - 0.5(-16.5\,\% - 10.75\,\%)} = -0.87.$$

Let $\mathscr{D}$ be the linear combination of deviation measures $\mathscr{D}_1, \ldots, \mathscr{D}_m$ defined by (2.1.2), and let $\mathscr{Q}_{\mathscr{D}}(r_M)$ be the set of risk identifiers for r_M with respect to $\mathscr{D}$. Then (8.2.16) implies

$$\beta_i = \frac{\mathrm{Cov}(-r_i, Q_M)}{\sum_{k=1}^{K} \lambda_k \mathscr{D}_k(r_M)}, \qquad Q_M \in \mathscr{Q}_{\mathscr{D}}(r_M). \tag{8.2.18}$$

Let $Q_M^{(1)}, \ldots, Q_M^{(m)}$ be risk identifiers of r_M with respect to $\mathscr{D}_1, \ldots, \mathscr{D}_m$, respectively, i.e. $Q_M^{(k)} \in \mathscr{Q}_{\mathscr{D}_k}(r_M) \subseteq \mathscr{Q}_k$, $k = 1, \ldots, m$, where $\mathscr{Q}_k$ is the risk envelope associated with $\mathscr{D}_k$, $k = 1, \ldots, m$. Then since the risk envelope $\mathscr{Q}$ associated with $\mathscr{D}$ is the closure of $\sum_{k=1}^{m} \lambda_k \mathscr{Q}_k$ [42, Proposition 4], we have $\sum_{k=1}^{m} \lambda_k Q_M^{(k)} \in \mathscr{Q}$. However, $\mathscr{D}_k(r_M) = E\left[\left(1 - Q_M^{(k)}\right) r_M\right]$, $k = 1, \ldots, m$, implies that

$$\mathscr{D}(r_M) = \sum_{k=1}^{m} \lambda_k \, E\left[\left(1 - Q_M^{(k)}\right) r_M\right] = E\left[\left(1 - \sum_{k=1}^{m} \lambda_k \, Q_M^{(k)}\right) r_M\right].$$

Consequently, $\sum_{k=1}^{m} \lambda_k Q_M^{(k)} \in \mathscr{Q}_{\mathscr{D}}(r_M) \subseteq \mathscr{Q}$, and (8.2.18) can be recast in the form

$$\beta_i = \frac{\sum_{k=1}^{m} \lambda_k \operatorname{Cov}\left(-r_i, Q_M^{(k)}\right)}{\sum_{k=1}^{m} \lambda_k \mathscr{D}_k(r_M)}, \qquad Q_M^{(k)} \in \mathscr{Q}_{\mathscr{D}_k}(r_M). \tag{8.2.19}$$

Example 8.5 (Mixed-CVaR deviation). For the mixed-CVaR deviation (2.1.4) with $0 < \alpha_1 < \cdots < \alpha_m < 1$, β_i is given by

$$\beta_i = \frac{\sum_{k=1}^{m} \lambda_k \operatorname{Cov}\left(-r_i, Q_M^{(k)}\right)}{\sum_{k=1}^{m} \lambda_k \mathrm{CVaR}_{\alpha_k}^{\Delta}(r_M)}, \qquad Q_M^{(k)} \in \mathscr{Q}_{\mathrm{CVaR}_{\alpha_k}^{\Delta}}(r_M). \tag{8.2.20}$$

In particular, if $\mathbb{P}\left[r_M = -\mathrm{VaR}_{\alpha_k}(r_M)\right] = 0$, $k = 1, \ldots, m$, then (8.2.20) takes the form

$$\beta_i = \frac{\sum_{k=1}^{m} \lambda_k \, E\left[E[r_i] - r_i \mid r_M \leq -\mathrm{VaR}_{\alpha_k}(r_M)\right]}{\sum_{k=1}^{m} \lambda_k \, E\left[E[r_M] - r_M \mid r_M \leq -\mathrm{VaR}_{\alpha_k}(r_M)\right]}.$$

In general, generalized betas may not be uniquely determined because either a master fund is not unique or $\mathscr{Q}_{\mathscr{D}}(r_M)$ is not a singleton. For generalized betas with other deviation measures, see [44].

Rockafellar et al. [45] proved that when investors' utility functions depend only on the mean and deviation of the portfolio rate of return and satisfy some additional conditions, a market equilibrium exists even if different groups of investors use different deviation measures. This result justifies the interpretation of the CAPM-like relations (8.2.15) in the sense of the classical CAPM (8.2.7) so that they can also serve as one-factor predictive models for expected rates of return of risky instruments.

8.2.4 CAPM Reinterpretation: Inverse Portfolio Problem

The generalized CAPM (8.2.15) with β_i given by (8.2.20) has the following interpretation. Suppose in (8.2.15), the master fund is of positive type and is known to have the rate of return r_M with $E[r_M] > r_0$, but the weights $\lambda_1, \ldots, \lambda_m$ in the mixed-CVaR deviation (2.1.4) are unknown. This is a so-called *inverse portfolio problem* for identifying investor's risk preferences: given a master fund, find investor's "risk profile" $\lambda_1, \ldots, \lambda_m$. Substituting (8.2.20) into (8.2.15), then multiplying (8.2.15) by $\mathscr{D}(r_M)$ and collecting terms with λ_k, we obtain a linear system for $\lambda_1, \ldots, \lambda_m$:

$$\begin{cases} \sum_{k=1}^{m} c_{ik}\, \lambda_k = 0, \quad i = 1, \ldots, n, \\ \sum_{k=1}^{m} \lambda_k = 1, \quad \lambda_k \geq 0, \quad k = 1, \ldots, m, \end{cases} \tag{8.2.21}$$

with

$$c_{ik} = \frac{E[r_i] - r_0}{E[r_M] - r_0} \mathrm{CVaR}_{\alpha_k}^{\Delta}(r_M) + \mathrm{Cov}\left(r_i, Q_M^{(k)}\right),$$

where $Q_M^{(k)}$ is a risk identifier of r_M corresponding to $\mathrm{CVaR}_{\alpha_k}^{\Delta}$. If r_M is discretely distributed: $\mathbb{P}[r_M = t_j] = p_j > 0$, $j = 1, \ldots, l$, with $\sum_{j=1}^{l} p_j = 1$, then $Q_M^{(k)} = (q_1, \ldots, q_l)$, where for each k, $q_1, \ldots, q_l$ are found from the optimization problem

$$\min_{q_1, \ldots, q_l} \sum_{j=1}^{l} q_j\, p_j\, t_j \quad \text{subject to} \quad \sum_{j=1}^{l} q_j\, p_j = 1, \quad 0 \leq q_j \leq \frac{1}{\alpha_k}, \quad j = 1, \ldots, l.$$

The system (8.2.21) has m unknowns, $n + 1$ linear equations, and m linear inequalities. However, since the master fund is a linear combination of the n risky instruments, the first n equations in (8.2.21) are linearly dependent. Consequently, if the $n + 1$ linear equations in (8.2.21) have n independent equations, the system (8.2.21) has a solution only if $m \geq n$ and may have infinitely many solutions if $m \geq n + 1$. If $m < n$, then an approximate solution to (8.2.21) can be obtained by the least squares method:

$$\min_{\lambda_1, \ldots, \lambda_m} \sum_{i=1}^{n-1} \left(\sum_{k=1}^{m} c_{ik}\, \lambda_k \right)^2 \quad \text{subject to} \quad \sum_{k=1}^{m} \lambda_k = 1, \quad \lambda_k \geq 0, \quad k = 1, \ldots, m,$$

see [25].

The inverse portfolio problem with an arbitrary deviation measure is solved in [15]. If it is known that a given rate of return X^* with $E[X^*] > 0$ is a solution of the Markowitz-type portfolio problem (8.2.10) with unknown deviation measure $\mathscr{D}$, then $\mathscr{D}$ can be represented through given X^* in the form [15, (25)]

$$\mathscr{D}(X) = E[X^*] \sup_{\alpha \in [0,1]} \frac{\mathrm{CVaR}_{\alpha}^{\Delta}(X)}{\mathrm{CVaR}_{\alpha}^{\Delta}(X^*)} \quad \text{for all } X \in \mathscr{L}^2(\Omega).$$

8.3 Deviation Minimization in Optimal Hedging

This section presents examples of hedging strategies with general deviation measures.

Example 8.6 (Mortgage pipeline hedging). Let a random variable r_0 model the volatile rate of a mortgage (loan), while the mortgage is in process (pipeline) (this is so-called mortgage pipeline risk), and let a random vector $r = (r_1, \dots, r_n)$ represent rates of return of n financial instruments available for hedging against the pipeline risk. A decision variable x_i is the portion of the mortgage amount invested into instrument i, $i = 1, \dots, n$, and the underperformance of the hedging portfolio $x = (x_1, \dots, x_n)$ with respect to the mortgage rate r_0 is given by the loss function $L(x, r_0, r) = r_0 - \sum_{i=1}^{n} r_i x_i \equiv r_0 - r^\top x$. The problem of optimal mortgage pipeline hedging is to minimize a deviation measure $\mathscr{D}$ of the loss:

$$\min_{x \in \mathbb{R}^n} \mathscr{D}\left(r_0 - r^\top x\right). \tag{8.3.1}$$

Observe that (8.3.1) imposes no additional constraints on the decision vector x, and for standard deviation, it has a closed-form solution provided that the covariance matrix Λ of $r_1, \dots, r_n$ is nonsingular:

$$x = \Lambda^{-1} \left(\mathrm{Cov}(r_0, r_1), \dots, \mathrm{Cov}(r_0, r_n)\right)^\top.$$

In Sect. 9.3, the case study "Mortgage Pipeline Hedging" implements the problem (8.3.1) in Portfolio Safeguard with standard deviation, mean absolute deviation (MAD), CVaR deviation, and two-tailed α-VaR deviation and solves the problem with real-life data.

Next example on optimal hedging is similar to Example 8.6, but in contrast to the latter, it imposes additional constraints on decision variables.

Example 8.7 (Optimal hedging of collateralized debt obligation (CDO)). Let a random variable θ_0 model the change in the uncertain value of a *CDO* book, and let random variables $\theta_1, \dots, \theta_n$ represent changes in uncertain spreads of n financial instruments (*credit default swaps* (CDS) and indices) available for hedging against the volatile CDO book value. In this case, a decision vector $x = (x_1, \dots, x_n)$ represents positions of the hedging instruments in a hedging portfolio, and the loss $L(x, \theta)$ is the underperformance of the hedging portfolio with respect to its target, which is a linear function of x and $\theta = (\theta_0, \theta_1, \dots, \theta_n)$. The CDO book hedging problem minimizes MAD of the loss $L(x, \theta)$ subject to budget and box constraints on the decision vector x:

$$\min_x \mathrm{MAD}(L(x, \theta)) \quad \text{subject to} \quad \sum_{i=1}^{n} c_i |x_i| \le C, \quad |x_i| \le k_i, \quad i = 1, \dots, n, \tag{8.3.2}$$

where $\sum_{i=1}^{n} c_i |x_i| \le C$ is the budget constraint with positive $c_1, \dots, c_n$, and $|x_i| \le k_i$, $i = 1, \dots, n$, are box constraints.

In Sect. 9.1, the case study "Optimal Hedging of CDO Book" implements the problem (8.3.2) in Portfolio Safeguard and solves the problem with real-life data.

8.4 Cash Matching with Deviation Constraints

Cash matching problems are a subclass of portfolio selection problems that aim to match existing cash obligations over some period of time with payments made from a portfolio. Let a fixed vector $p \in \mathbb{R}^n$ represent initial prices of n bonds, and let a vector $x \in \mathbb{R}^n$ represent portfolio positions (weights) in those bonds. Then the portfolio initial price is given by $p^\top x$. Let $L(x,\theta)$ be a loss function in matching cash obligations, where θ is a random vector. It is assumed that $L(x,\theta)$ is a linear function of x. Typically, the portfolio is imposed a *self-financing* constraint $E[L(x,\theta)] = 0$, and shorting is not allowed: $x \geq 0$.

Example 8.8 (Cash matching bond portfolio problem I). This cash matching problem minimizes the initial cost of a bond portfolio subject to the self-financing constraint, a constraint on the MAD of the loss function, and subject to the no-shorting constraint:

$$\min_x \; p^\top x \quad \text{subject to} \quad \mathrm{MAD}(L(x,\theta)) \leq w, \quad E[L(x,\theta)] = 0, \quad x \geq 0, \tag{8.4.1}$$

where $w \in \mathbb{R}^+$ is a given fixed value. Observe that the constraint $\mathrm{MAD}(L(x,\theta)) \leq w$ may render the problem infeasible.

Example 8.9 (Cash matching bond portfolio problem II). The following formulation avoids possible infeasibility in the problem (8.4.1) by introducing an auxiliary variable z and a penalty coefficient $K \in \mathbb{R}^+$:

$$\begin{aligned} &\min_{x,z} \; p^\top x + K\,z \\ &\text{subject to} \;\; \mathrm{MAD}(L(x,\theta)) - z \leq w, \\ &\qquad\qquad\quad E[L(x,\theta)] = 0, \quad z \geq 0, \quad x \geq 0. \end{aligned} \tag{8.4.2}$$

In Sect. 9.4, the case study "Cash Matching Bond Portfolio" implements the problems (8.4.1) and (8.4.2) in Portfolio Safeguard and solves them with real-life data.

8.5 Risk Minimization in Portfolio Theory

A typical portfolio risk minimization problem is similar to the Markowitz-type portfolio selection (8.2.10) and is formulated as follows. Let a random vector $r : \Omega \to \mathbb{R}^n$ represent rates of return of n risky instruments, and let a vector $x = (x_1, \ldots, x_n) \in \mathbb{R}^n$ represent portfolio positions (weights) in those instruments.

Then $X = r^\top x$ is the portfolio rate of return, and given a risk measure $\mathscr{R}$, the problem is to *minimize the portfolio risk* $\mathscr{R}(X)$ *subject to the budget constraint* $e^\top x = 1$, *a constraint on the portfolio expected rate of return* $E[X] \geq \mu$ *with specified* μ *and subject to other "technical" constraints, e.g., box constraints on portfolio positions*:

$$\min_{X \in \mathscr{X}} \mathscr{R}(X) \quad \text{subject to} \quad E[X] \geq \mu$$
$$\text{with } \mathscr{X} = \left\{X = r^\top x \;\middle|\; e^\top x = 1,\ \ x \in \mathscr{C}\right\}, \tag{8.5.1}$$

where the set $\mathscr{C}$ incorporates all other constraints on x. Observe that the risk measure $\mathscr{R}$ need not necessarily be coherent, risk averse, or even convex, though the use of nonconvex functionals poses certain optimization difficulties.

Example 8.10 (VaR and CVaR minimization). In credit-risk management,[3] portfolio problems minimize either VaR or CVaR of the portfolio rate of return $X = r^\top x$ subject to a constraint on portfolio expected rate of return and subject to budget and box constraints on the portfolio positions:

$$\min_{x} \ \mathrm{VaR}_\alpha\left(r^\top x\right) \quad \text{subject to} \quad E[r^\top x] \geq \mu, \quad e^\top x = 1, \quad 0 \leq x \leq 2x^0, \tag{8.5.2}$$

$$\min_{x} \ \mathrm{CVaR}_\alpha\left(r^\top x\right) \quad \text{subject to} \quad E[r^\top x] \geq \mu, \quad e^\top x = 1, \quad 0 \leq x \leq 2x^0, \tag{8.5.3}$$

where x^0 is the nonnegative n-dimensional vector of positions of a given initial portfolio; see [21] for details.

In Sect. 9.15, the case study "Portfolio Credit-Risk Optimization Modeled by Scenarios and Mixtures of Normal Distributions" implements the problems (8.5.2) and (8.5.3) in Portfolio Safeguard and solves them with real-life data.

VaR and CVaR for Mixtures of Normal Distributions

Two popular approaches to modeling of distributions of random variables of interest, e.g., portfolio rate of return X, are "conventional" scenario approach and the one that uses mixtures of independent normal distributions; see [21]. (The case study in Sect. 9.15 models the distribution of X in the problems (8.5.2) and (8.5.3) with both approaches.) This subsection presents analytical formulae for VaR and CVaR of a continuously distributed random variable whose PDF is given by a mixture of normal distributions.

---

[3] *Credit risk* also known as *default risk* is associated with a potential failure (default) of a borrower to either repay credit or meet certain financial obligations.

Suppose there are m scenarios representing states of the economy with positive probabilities $p_1, \ldots, p_m$ summing into 1. Let the rate of return of instrument i under scenario j be a normally distributed random variable r_{ij} with mean μ_{ij} and variance σ_{ij}^2, i.e., $r_{ij} \sim N(\mu_{ij}, \sigma_{ij}^2)$. It is assumed that $r_{1j}, \ldots, r_{nj}$ are independent for each j. Consequently, under scenario j, the portfolio rate of return X is $\sum_{j=1}^n r_{ij} x_j$, which is normally distributed with mean $\mu_j(x) = \sum_{i=1}^n \mu_{ij}\, x_i$ and variance $\sigma_j^2(x) = \sum_{i=1}^n \sigma_{ij}^2\, x_i^2$.

Now across scenarios, the portfolio rate of return is modeled as a lottery with outcomes being random variables with normal distributions $N(\mu_1(x), \sigma_1^2(x)), \ldots, N(\mu_m(x), \sigma_m^2(x))$ and having probabilities $p_1, \ldots, p_m$. In other words, X is a random variable, whose distribution is a mixture of the normal distributions with the PDF and CDF determined by

$$f_X(t) = \sum_{j=1}^m \frac{p_j}{\sigma_j(x)}\, \phi\left(\frac{t - \mu_j(x)}{\sigma_j(x)}\right), \qquad F_X(t) = \sum_{j=1}^m p_j\, \Phi\left(\frac{t - \mu_j(x)}{\sigma_j(x)}\right),$$

respectively, where $\phi(t) = \mathrm{e}^{-t^2/2}/\sqrt{2\pi}$ and $\Phi(t) = \int_{-\infty}^t \phi(\tau)\, d\tau$ are the PDF and CDF of the standard normal distribution $N(0,1)$, respectively.

Let $F_X^{-1}(\alpha)$, $\alpha \in (0,1)$, be the inverse of $F_X(t)$. Then

$$\mathrm{VaR}_\alpha(X) = -F_X^{-1}(\alpha), \tag{8.5.4}$$

and

$$\begin{aligned} \mathrm{CVaR}_\alpha(X) &= -\frac{1}{\alpha}\int_0^\alpha F_X^{-1}(s)\, ds = -\frac{1}{\alpha}\int_{-\infty}^{F_X^{-1}(\alpha)} t\, f_X(t)\, dt \\ &= \frac{1}{\alpha}\sum_{j=1}^m p_j \left(\sigma_j(x)\phi(t) - \mu_j(x)\Phi(t)\right)\Bigg|_{t=\frac{F_X^{-1}(\alpha)-\mu_j(x)}{\sigma_j(x)}}. \end{aligned} \tag{8.5.5}$$

For $m = 1$, (8.5.4) and (8.5.5) simplify to

$$\mathrm{VaR}_\alpha(X) = -\mu_1(x) - \sigma_1(x)\, \Phi^{-1}(\alpha)$$

and

$$\mathrm{CVaR}_\alpha(X) = -\mu_1(x) + \frac{\sigma_1(x)}{\alpha}\, \phi\left(\Phi^{-1}(\alpha)\right),$$

respectively, which are well-known formulae for VaR and CVaR for a normally distributed random variable X with mean $\mu_1(x)$ and standard deviation $\sigma_1(x)$.

Optimization

Though, closed-form representations (8.5.4) and (8.5.5) are suitable for analytical treatment, they are hard to optimize with respect to x. Instead, we can use the relationships $\mathrm{VaR}_\alpha(X) = -q_X^+(\alpha)$ and $\mathrm{CVaR}_\alpha(X) = -\overline{q}_X(\alpha)$ and the formula (1.4.4) to evaluate VaR and CVaR by

$$\mathrm{VaR}_\alpha(X) = \min_\tau\ (-\tau) \quad \text{subject to} \quad \sum_{j=1}^m p_j\, \Phi\left(\frac{\tau - \mu_j(x)}{\sigma_j(x)}\right) \le \alpha, \tag{8.5.6}$$

$$\mathrm{CVaR}_\alpha(X) = \min_\tau \left(-\tau + \frac{1}{\alpha}\sum_{j=1}^m p_j\, \sigma_j(x)\, \Psi\left(\frac{\tau - \mu_j(x)}{\sigma_j(x)}\right)\right), \tag{8.5.7}$$

where $\Psi(t) = t\,\Phi(t) + \phi(t)$.

Now suppose that the distribution of the portfolio rate of return X in Example 8.10 is modeled by the mixture of normal distributions as discussed above, then the portfolio problems (8.5.2) and (8.5.3) can be reformulated with the formulae (8.5.6) and (8.5.7) in the form

$$\begin{aligned} &\min_{x,\tau}\ (-\tau) \\ &\text{subject to} \quad \sum_{j=1}^m p_j\, \Phi\left(\frac{\tau - \mu_j(x)}{\sigma_j(x)}\right) \le \alpha, \\ &\qquad\qquad E[r^\top x] \ge \mu, \quad e^\top x = 1, \quad 0 \le x \le 2x^0, \end{aligned} \tag{8.5.8}$$

and

$$\begin{aligned} &\min_{x,\tau} \quad -\tau + \frac{1}{\alpha}\sum_{j=1}^m p_j\, \sigma_j(x)\, \Psi\left(\frac{\tau - \mu_j(x)}{\sigma_j(x)}\right) \\ &\text{subject to} \quad E[r^\top x] \ge \mu, \quad e^\top x = 1, \quad 0 \le x \le 2x^0, \end{aligned} \tag{8.5.9}$$

respectively.

In Sect. 9.15, the case study "Portfolio Credit-Risk Optimization Modeled by Scenarios and Mixtures of Normal Distributions" implements the problems (8.5.8) and (8.5.9) in Portfolio Safeguard and solves the problems with real-life data.

8.6 Applications with Downside Risk Constraints

This section demonstrates optimization problems with constraints on the downside risk of not meeting target in hedging, cash flow matching, and test selection.

In some applications, it is customary to conduct analysis in terms of "underperformance" and "loss," e.g., in terms of negative portfolio rate of return. Given a random loss L, CVaR of L with $\alpha \in [0, 1)$ can be redefined as *the average of the largest* $(1-\alpha) \cdot 100\,\%$ *outcomes of* L. To distinguish it from the original CVaR, we call it *CVaR for loss* and denote it by CVaR_α^-. The mathematical definition of CVaR_α^- is similar to that of CVaR_α, namely,

$$\mathrm{CVaR}_\alpha^-(L) = \frac{1}{1-\alpha}\int_\alpha^1 q_L^-(s)\,ds, \tag{8.6.1}$$

where $q_L^-(s)$ is the lower s-quantile of L.[4] In fact, if loss L is merely a random variable X with changed sign, i.e., $L = -X$, then $\mathrm{CVaR}_\alpha(X)$ and $\mathrm{CVaR}_\alpha^-(L)$ are related by

$$\mathrm{CVaR}_\alpha^-(L) = \mathrm{CVaR}_{1-\alpha}(X) \quad \text{for} \quad L = -X, \tag{8.6.2}$$

so that all optimization formulae for CVaR_α can be readily translated for CVaR_α^-.

Example 8.11 (Hedging portfolio of options with CVaR constraint). Suppose a portfolio of n options is to be formed at time $t = 0$ to hedge against a target portfolio at expiration time $t = T$. A decision vector $x \in \mathbb{R}^n$ represents positions of the hedging portfolio in the options. Let $P(x)$ be the price of the portfolio at $t = 0$, and let $L(x, \theta)$ be underperformance (loss) of the hedging portfolio at the final time T, where θ is a random vector. Both $P(x)$ and $L(x, \theta)$ are assumed to be linear functions of x. The hedging portfolio problem minimizes $P(x)$ subject to a constraint on the average of the largest $(1-\alpha) \cdot 100\,\%$ losses not to exceed a given threshold d and subject to box constraints on x:

$$\min_x\ P(x) \quad \text{subject to} \quad \mathrm{CVaR}_\alpha^-(L(x,\theta)) \le d, \quad l \le x \le u, \tag{8.6.3}$$

where $l \in \mathbb{R}^n$ and $u \in \mathbb{R}^n$ are given vectors of bounds.

In Sect. 9.2, the case study "Hedging Portfolio of Options" implements the problem (8.6.3) in Portfolio Safeguard and solves the problem with real-life data.

Example 8.12 (Cash flow matching with CVaR constraint). Suppose $l_1, \dots, l_T$ are financial liabilities to be met at time moments $1, \dots, T$, and suppose there are n bonds available for investment at each time. Let p_{tk} be uncertain price of bond k at time t, and let c_{jtk} be uncertain payment at time t from bond k purchased at time $j < t$. In this case, decision variable x_{tk} is the number of shares of bond k purchased at time t, and $L_t = l_t + p_t^\top x_t - \sum_{j=0}^{t-1} c_{jt}^\top x_j$ is the underperformance of a replicating portfolio versus the liability l_t at t, where $p_t = (p_{t1}, \dots, p_{tn})$, $c_{jt} = (c_{jt1}, \dots, c_{jtn})$, and $x_j = (x_{j1}, \dots, x_{jn})$. The loss is defined as the

---

[4] In [39, 40], CVaR was originally introduced by (8.6.1); however, in later works [41–44, 46], it was reintroduced as (2.0.2).

worst underperformance over the period $[1,T]$: $L = \max_{1\le t\le T} L_t$. The cash flow matching problem minimizes the portfolio cost at $t = 0$ subject to a constraint that the average of the largest $(1-\alpha)\cdot 100\,\%$ losses is not positive:

$$\min_{x_1,\dots,x_T} \quad p_0^\top x_0 \quad \text{subject to} \quad \mathrm{CVaR}_\alpha^-(L) \le 0, \quad x_t \ge 0, \quad t = 1,\dots,T, \qquad (8.6.4)$$

where the constraint $x_t \ge 0$ excludes shorting.

In Portfolio Safeguard, CVaR$^-$ of the maximum of linear functions in Example 8.12 is given by the function cvar_max_risk.

In Sect. 9.5, the case study "Cash Flow Matching with CVaR Constraints" implements the problem (8.6.4) in Portfolio Safeguard and solves the problem with real-life data.

Optimal Test Selection

Suppose there are n different tests with values $c_1 \in \mathbb{R}_0^+,\dots,c_n \in \mathbb{R}_0^+$, and suppose there are m different resources with finite capacities $b_1 \in \mathbb{R}^+,\dots,b_m \in \mathbb{R}^+$. Decision variable x_i is 1 if test i is conducted and is zero otherwise, so that $x = (x_1,\dots,x_n)$ is the boolean decision vector (no partial test is allowed). Next two examples present different approaches to optimal test selection.

Example 8.13 (Robust test selection). Suppose, on average, test i consumes amount $a_{ij} \in \mathbb{R}_0^+$ of resource j. Let $L_j(x)$ be interpreted as a discrete random variable which takes on values $a_{1j}x_1,\dots,a_{nj}x_n$ with equal probabilities $1/n$, then the average of the (average) consumption of resource j by its $(1-\alpha)\cdot 100\,\%$ largest consumers can be determined by

$$\mathrm{CVaR}_\alpha^-(L_j(x)) = \max_{q_1,\dots,q_n} \sum_{i=1}^n a_{ij}x_i\,q_i \ \text{ subject to } \sum_{i=1}^n q_i = 1,\ 0 \le q_i \le \frac{1}{(1-\alpha)\,n}.$$

This formula for CVaR$^-$ is similar to (1.4.6). It computes the average of the right $(1-\alpha)$-tail of $L_j(x)$, whereas (1.4.6) computes the average of the left α-tail of X.

To guarantee that the constraints on resource consumptions are met not only on average, the average consumption of resource j, i.e., $\sum_{i=1}^n a_{ij}x_i$, is increased by $\lambda\cdot 100\,\%$ of $\mathrm{CVaR}_\alpha(L_j(x))$. Thus, the robust test selection problem maximizes the value of selected tests subject to constraints on resources:

$$\begin{aligned} &\max_x \ \sum_{i=1}^n c_i x_i \\ &\text{subject to} \ \sum_{i=1}^n a_{ij}x_i + \lambda\,\mathrm{CVaR}_\alpha^-(L_j(x)) \le b_j, \quad j = 1,\dots,m, \\ &\qquad x_i \in \{0,1\}, \quad i = 1,\dots,n. \end{aligned} \qquad (8.6.5)$$

Example 8.14 (Test selection with chance constraint). This time, consumption of resource j by test i is assumed to be a random variable θ_{ij}. Let $R_j(x) = b_j - \sum_{i=1}^{n} \theta_{ij} x_i$ be the uncertain remainder of resource j. The optimal test selection problem then maximizes the value of selected tests subject to a constraint that the probability to deplete at least one resource is less than or equal to d:

$$\begin{aligned} &\max_{x} \sum_{i=1}^{n} c_i x_i \\ &\text{subject to } 1 - \mathbb{P}[R_1(x) \geq 0, \ldots, R_n(x) \geq 0] \leq d, \\ &\qquad x_i \in \{0, 1\}, \quad i = 1, \ldots, n. \end{aligned} \tag{8.6.6}$$

In Sect. 9.10, the case study "Optimal Test Selection" implements the problems (8.6.5) and (8.6.6) in Portfolio Safeguard and solves them with real-life data.

8.7 Portfolio Optimization with Drawdown Measure

For a long investment period, active portfolio management [or commodity trading advisor (CTA)] considers not only terminal values of risk and expectation of portfolio rate of return, it also imposes several restrictions on portfolio *drawdown* during the investment period. Usually, investors quit an investment fund after either a single large drawdown (greater than 20 %) or a small but prolonged drawdown (over a year); see [6, 7]. Portfolio drawdown can be efficiently managed in the optimization framework discussed in this chapter.

8.7.1 Drawdown Measure

We begin with an abstract definition of drawdown. Let $\xi(t)$ be an arbitrary real-valued bounded function on interval $[0, a]$ $(a > 0)$. Drawdown of ξ at time t with τ-window $(\tau > 0)$ is defined by

$$d_t = \max_{t_\tau \leq \ell \leq t} \xi(\ell) - \xi(t), \qquad t_\tau = \max\{t - \tau, 0\}, \quad t \in [0, a],$$

which is the drop of $\xi(t)$ from its current maximum on the interval $[t_\tau, t]$. Negative drawdown is often referred to as "underwater curve."

Usually, t represents time, and the function $\xi(t)$ is specified only at T time moments $t = 1, \ldots, T$: $\xi(t) = \xi_t$. The sequence $\{\xi_t\}_{t=1}^{T}$ is regarded as a sample path of ξ. Suppose there are S sample paths $\{\xi_{st}\}_{t=1}^{T}$, $s = 1, \ldots, S$, of ξ for the same time moments, and suppose sample path s has probability $p_s > 0$,

$s = 1, \ldots, S$ $\left(\sum_{s=1}^{S} p_s = 1\right)$. Then on sample path s, the drawdown of ξ with τ-window is defined by

$$d_{st} = \max_{t_\tau \le k \le t} \xi_{sk} - \xi_{st}, \qquad t_\tau = \max\{t - \tau, 1\}, \tag{8.7.1}$$

where $t = 1, \ldots, T$ and τ is a fixed integer between 1 and T.

We consider three frequently used drawdown measures: *maximum drawdown*, *average drawdown*, and *conditional drawdown-at-risk (CDaR)*. Maximum and average drawdowns are defined by

$$\text{MaxDD}(\xi) = \max_{\substack{1 \le s \le S,\\ 1 \le t \le T}} d_{st}$$

and

$$\text{AvgDD}(\xi) = \frac{1}{T} \sum_{s=1}^{S} \sum_{t=1}^{T} p_s\, d_{st},$$

respectively, whereas for a given $\alpha \in [0, 1)$, *CDaR is the average of* $(1 - \alpha) \cdot 100\,\%$ *largest drawdowns of the set* $\{d_{st} \,|\, t = 1, \ldots, T,\ s = 1, \ldots, S\}$ and is formally determined as

$$\text{CDaR}_\alpha(\xi) = \max_{\{q_{st}\} \in Q} \sum_{s=1}^{S} \sum_{t=1}^{T} p_s\, q_{st}\, d_{st} \tag{8.7.2}$$

with the set Q given by

$$Q = \left\{ \{q_{st}\}_{s,t=1}^{S,T} \,\middle|\, \sum_{s=1}^{S} \sum_{t=1}^{T} p_s q_{st} = 1,\ 0 \le q_{st} \le \frac{1}{(1-\alpha)T}, \right.$$
$$\left. s = 1, \ldots, S,\ t = 1, \ldots, T \right\}.$$

For $\alpha = 1$, $\text{CDaR}_\alpha(\xi)$ has the form of (8.7.2) with the constraint $0 \le q_{st} \le 1/((1-\alpha)T)$ in Q replaced by $q_{st} \ge 0$ for all s and t. When $(1 - \alpha)\, T\, S$ is an integer, the representation (8.7.2) simplifies to

$$\text{CDaR}_\alpha(\xi) = \sum_{s=1}^{S} \sum_{t=1}^{T} q_{st}^*\, p_s\, d_{st}, \tag{8.7.3}$$

where $q_{st}^* = 1/((1-\alpha)T)$ if d_{st} is one of the $(1 - \alpha)\, T\, S$ largest drawdowns and $q_{st}^* = 0$ otherwise.

In fact, the maximum drawdown and the average drawdown are CDaR for $\alpha = 1$ and $\alpha = 0$, respectively:

$$\mathrm{MaxDD}(\xi) = \lim_{\alpha \to 1-} \mathrm{CDaR}_\alpha(\xi), \qquad \mathrm{AvgDD}(\xi) = \mathrm{CDaR}_0(\xi).$$

In Portfolio Safeguard, $\mathrm{MaxDD}(\xi)$, $\mathrm{AvgDD}(\xi)$, and $\mathrm{CDaR}_\alpha(\xi)$ are implemented by the functions drawdown_dev_max, drawdown_dev_avg, and cdar_dev for a single sample path and by the functions drawdownmulti_dev_max, drawdownmulti_dev_avg, and cdarmulti_dev for multiple sample paths.

8.7.2 Linear Programming Representations

With a formula similar to (1.4.4), the definition (8.7.2) can be restated as

$$\mathrm{CDaR}_\alpha(\xi) = \min_{c \in \mathbb{R}} \left(c + \frac{1}{(1-\alpha)T} \sum_{s=1}^{S} \sum_{t=1}^{T} p_s \, [d_{st} - c]_+ \right),$$

which yields the representation of CDaR in the form of linear program

$$\begin{aligned} \mathrm{CDaR}_\alpha(\xi) = \quad & \min_{c,\, y_{st},\, z_{st}} \left(c + \frac{1}{(1-\alpha)T} \sum_{s=1}^{S} \sum_{t=1}^{T} p_s y_{st} \right) \\ & \text{subject to} \\ & \left. \begin{aligned} & z_{st} \geq z_{s,t-1} + \xi_{s,t-1} - \xi_{st} \\ & y_{st} \geq z_{st} - c \\ & y_{st} \geq 0, \quad z_{st} \geq 0 \\ & \xi_{s0} = 0, \quad z_{s0} = 0 \end{aligned} \right\} \quad \text{for all } s \text{ and } t. \end{aligned} \tag{8.7.4}$$

For the maximum and average drawdowns, corresponding representations (linear programs) are given by

$$\begin{aligned} \mathrm{MaxDD}(\xi) = \quad & \min_{y,\, z_{st}} \; y \\ & \text{subject to} \\ & \left. \begin{aligned} & z_{st} \geq z_{s,t-1} + \xi_{s,t-1} - \xi_{st} \\ & y \geq z_{st}, \quad z_{st} \geq 0 \\ & \xi_{s0} = 0, \quad z_{s0} = 0 \end{aligned} \right\} \quad \text{for all } s \text{ and } t \end{aligned} \tag{8.7.5}$$

and

$$\mathrm{AvgDD}(\xi) = \min_{y_{st},\, z_{st}} \frac{1}{T} \sum_{s=1}^{S} \sum_{t=1}^{T} p_s y_{st}$$

$$\text{subject to} \qquad (8.7.6)$$

$$\left. \begin{array}{l} z_{st} \geq z_{s,t-1} + \xi_{s,t-1} - \xi_{st} \\ y_{st} \geq z_{st}, \quad z_{st} \geq 0 \\ \xi_{s0} = 0, \quad z_{s0} = 0 \end{array} \right\} \quad \text{for all } s \text{ and } t.$$

The linear programs (8.7.4)–(8.7.6) are useful for evaluating CDaR for relatively small S and T. However, Portfolio Safeguard has specially designed nonlinear programming subroutines (not based on linear programming) for evaluating and optimizing CDaR, which are more efficient.

8.7.3 *Portfolio Problem Formulations*

Suppose there are n risky instruments available on the market, and suppose there is a *zero-rate* risk-free instrument ($r_0 = 0$). An investment problem is planned for T consecutive time periods $t = 1, \ldots, T$, over which the rates of return of the risky instruments are modeled by S sample paths: $r_{s1}, \ldots, r_{sT}$, $s = 1, \ldots, S$, where $r_{st} = \left(r_{st}^1, \ldots, r_{st}^n\right)$ with r_{st}^i being the rate of return of instrument i for $i = 1, \ldots, n$, $s = 1, \ldots, S$, $t = 1, \ldots, T$. Sample path s has the probability of occurrence $p_s > 0$, where $\sum_{s=1}^{S} p_s = 1$. Let w_{st}^i denote the cumulative rate of return of instrument i in sample path s at time t, which can be either *uncompounded cumulative rate of return*, defined by $w_{st}^i = \sum_{j=1}^{t} r_{sj}^i$, or *compounded cumulative rate of return*, defined by $w_{st}^i = \prod_{j=1}^{t} \left(1 + r_{sj}^i\right) - 1$. Further, w_{st}^i will be referred simply to as cumulative rate of return, and all results will hold for both definitions. For any sample path and time moment, the zero-rate risk-free instrument has zero compounded and uncompounded cumulative rates of return: $w_{st}^0 = 0$, $s = 1, \ldots, S$, $t = 1, \ldots, T$. Next two examples illustrate portfolio settings with compounded and uncompounded cumulative rates of return.

Example 8.15 (Portfolio with compounded cumulative rate of return). In this example, a portfolio follows a traditional setting, where capital C is invested in the risk-free instrument and in the n risky instruments (e.g., stocks and bonds) with capital weights x_0 and $x_1, \ldots, x_n$, respectively. Note that x_i can be negative, which corresponds to shorting, and the weights sum up to one: $x_0 + \sum_{i=1}^{n} x_i = 1$ (budget constraint). At the end of time period t in sample path s, instrument i accumulates $(1+w_{st}^i)\, x_i\, C$, and the portfolio has the value $C\left[(1+w_{st}^0)\, x_0 + \sum_{i=1}^{n} (1+w_{st}^i)\, x_i\right] \equiv C\left[1 + \sum_{i=1}^{n} w_{st}^i\, x_i\right]$. Consequently, for sample path s at time t, the portfolio cumulative rate of return is given by

$$w_{st}^{p}(x) = \sum_{i=1}^{n} x_i \, w_{st}^{i}, \tag{8.7.7}$$

and the corresponding portfolio drawdown with τ-window ($\tau \in \{1,\ldots,T\}$) is defined by

$$d_{st}^{p}(x) = \max_{t_\tau \le k \le t} w_{sk}^{p}(x) - w_{st}^{p}(x), \qquad t = 1,\ldots,T, \tag{8.7.8}$$

where $t_\tau = \max\{t - \tau, 0\}$. Observe that since the risk-free instrument has zero rate of return, the portfolio rate of return (8.7.7) does not involve the risk-free instrument. Moreover, x_0 can be expressed from the budget constraint as $x_0 = 1 - \sum_{i=1}^{n} x_i$, and consequently, a portfolio problem can be formulated only in terms of $x = (x_1,\ldots,x_n)$ with no budget constraint on x. Also, it is worth mentioning that $x_1,\ldots,x_n$ have the meaning of portfolio weights only at the initial time moment, because the proportions of the portfolio capital in each instrument may change in time, and $x_1,\ldots,x_n$ may not coincide with the true portfolio weights at times $t = 1,\ldots,T$.

Example 8.16 (Portfolio with uncompounded cumulative rate of return). This setting is typical for a portfolio with positions in futures and other derivatives. Suppose a portfolio has positions in n futures contracts. In this case, initial capital C is not invested, but is used rather as a deposit, and x_i is a position (positive or negative) opened in instrument i. Also, there is no position in the zero-rate risk-free instrument, and the sum of the positions $x_1,\ldots,x_n$ is not constrained. The profit (or loss) that the portfolio $x = (x_1,\ldots,x_n)$ makes over time period t in sample path s, i.e., $\sum_{i=1}^{n} r_{st}^{i}\, x_i$, is transferred from the portfolio to a separate account, so that in the beginning of period $t+1$, the positions $x_1,\ldots,x_n$ remain the same. Then, over t periods since investment, the account accumulates *uncompounded cumulative profit* $\sum_{k=1}^{t} \left(\sum_{i=1}^{n} r_{sk}^{i}\, x_i\right) = \sum_{i=1}^{n} \left(\sum_{k=1}^{t} r_{sk}^{i}\right) x_i = \sum_{i=1}^{n} w_{st}^{i}\, x_i = w_{st}^{p}(x)$. In other words, the profit of the futures contract portfolio is mathematically identical to the cumulative rate of return (8.7.7), although $x_1,\ldots,x_n$ here and in (8.7.7) have different meaning. Consequently, the drawdown for the futures contract portfolio is defined by (8.7.8) with $w_{st}^{p}(x)$ being the portfolio uncompounded cumulative profit.

Further, we will not distinguish portfolios with compounded and uncompounded cumulative rates of return.

A portfolio optimization problem with CDaR can be formulated either as minimizing the portfolio CDaR over T time periods subject to a constraint on the portfolio expected rate of return at time T:

$$\min_{x} \; \mathrm{CDaR}_\alpha\left(w^{p}(x)\right) \quad \text{subject to} \quad \sum_{s=1}^{S} p_s w_{sT}^{p}(x) \ge \Delta, \tag{8.7.9}$$

or as maximizing the portfolio expected rate of return at time T subject to a constraint on the portfolio CDaR over T time periods:

$$\max_{x} \sum_{s=1}^{S} p_s w_{sT}^p(x) \quad \text{subject to} \quad \mathrm{CDaR}_\alpha\left(w^p(x)\right) \leq \nu, \tag{8.7.10}$$

where $x = (x_1, \ldots, x_n)$ is the vector of portfolio weights, $\Delta > 0$, $\nu > 0$, and $\mathrm{CDaR}_\alpha\left(w^p(x)\right)$ is a convex and positively homogeneous function of x (see [7]). The problem (8.7.9) is similar to Markowitz-type portfolio selection (8.2.10), in which the deviation measure $\mathscr{D}$ is replaced by portfolio CDaR, whereas the problem (8.7.10) is used in active portfolio management, where $w^p(x)$ is the portfolio profit; see [6]. The problems (8.7.9) and (8.7.10) are both convex and are equivalent in the sense that by varying Δ in (8.7.9) and ν in (8.7.10), the same set of optimal solutions (called efficient frontier) will be generated.

The case studies in Sects. 9.19 and 9.20 solve the problem (8.7.9) for single-path and multiple-path datasets subject to additional box constraints on portfolio weights, $x_{\min} \leq x_i \leq x_{\max}$, $i = 1, \ldots, n$, whereas the case study in Sect. 9.21 compares solutions to the problem (8.7.9) for multiple paths and for the single path obtained by aggregating those multiple paths. (Portfolio Safeguard does not reduce (8.7.9) to linear programming.)

8.7.4 CAPM with CDaR

The portfolio problems (8.7.9) and (8.7.10) lead to the same necessary optimality conditions for portfolio weights, which can be stated in the form of CAPM. It is assumed that the constraint $\sum_{s=1}^{S} p_s w_{sT}^p(x) \geq \Delta$ is feasible and that $\mathrm{CDaR}_\alpha\left(w^p(x)\right) > 0$ for any $x \neq 0$. Next theorem is proved in [63].

Theorem 8.3 (CAPM with CDaR for multiple sample paths and arbitrary α). *Let $x^* = (x_1^*, \ldots, x_n^*)$ be an optimal solution to either (8.7.9) or (8.7.10), and let $w^M = w^p(x^*) \equiv \left\{w_{st}^p\right\}_{s,t=1}^{S,T}$ be the cumulative rates of return of the optimal portfolio; then the necessary optimality conditions for x^* can be restated in the form of CAPM*

$$\sum_{s=1}^{S} p_s w_{sT}^i = \beta_{\mathrm{DD}}^i \sum_{s=1}^{S} p_s\, w_{sT}^M, \quad \beta_{\mathrm{DD}}^i = \frac{\sum_{s=1}^{S}\sum_{t=1}^{T} p_s\, q_{st}^*\left(w_{s,k^*(s,t)}^i - w_{st}^i\right)}{\mathrm{CDaR}_\alpha\left(w^M\right)} \tag{8.7.11}$$

for $i = 1, \ldots, n$, where β_{DD}^i is the drawdown beta of instrument i, $\{q_{st}^*\}_{s,t=1}^{S,T}$ is an element from Q at which maximum is attained in (8.7.2) for $w^p(x^*)$, $k^*(s,t) \in \operatorname{argmax}_{t_\tau \leq k \leq t} w_{sk}^p(x^*)$, and $\mathrm{CDaR}_\alpha(w^M) \neq 0$ since $\mathrm{CDaR}_\alpha\left(w^p(x)\right)$ is assumed to be positive for any $x \neq 0$ and the constraint $\sum_{s=1}^{S} p_s w_{sT}^p(x) \geq \Delta$ precludes x to be 0.

If in Theorem 8.3, $S = 1$, $(1-\alpha)T$ is an integer, and $k^*(1,t)$ and $\{q^*_{1t}\}_{t=1}^T$ are *uniquely* determined, the CAPM (8.7.11) simplifies.

Corollary 8.3 (CAPM with CDaR for a single sample-path when $(1-\alpha)T$ is integer). *For a single sample-path ($S = 1$), $p_1 = 1$ and the index (or variable) "s" in all entries of* (8.7.11) *is dropped. In this case, the CAPM* (8.7.11) *takes the form*

$$w^i_T = \beta^i_{\rm DD} \cdot w^M_T \qquad \textit{with} \qquad \beta^i_{\rm DD} = \frac{\sum_{t=1}^T q^*_t (w^i_{k^*(t)} - w^i_t)}{\mathrm{CDaR}_\alpha(w^M)}. \tag{8.7.12}$$

For integer $(1-\alpha)T$, $q^*_t = 1/((1-\alpha)T)$ if d^M_t is one of the $(1-\alpha)T$ largest drawdowns of $d^M_1, \ldots, d^M_T$ of the optimal portfolio on the single sample path and $q^*_t = 0$ otherwise. In (8.7.12), $\mathrm{CDaR}_\alpha(w^M) \neq 0$ *as explained in Theorem 8.3, and it is assumed that q^*_t and $k^*(t)$ are uniquely determined for all $t = 1, \ldots, T$.*

The numerator in $\beta^i_{\rm DD}$ in (8.7.12) is the average rate of return of instrument i over time periods corresponding to the $(1-\alpha)T$ largest drawdowns of the optimal portfolio, where $w^i_t - w^i_{k^*(t)}$ is the cumulative rate of return of instrument i from the optimal portfolio peak time $k^*(t)$ to time t.

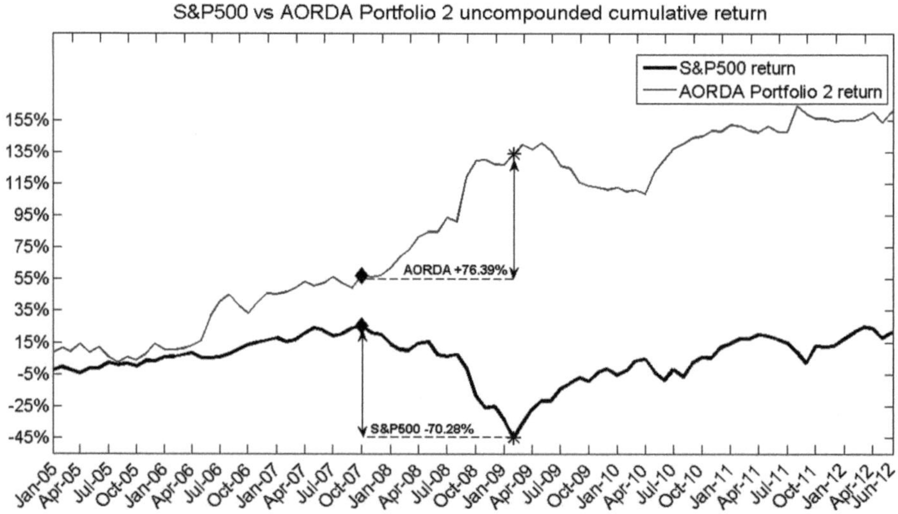

Fig. 8.3 Uncompounded monthly cumulative rates of return of the S&P500 index and of the AORDA Portfolio 2 ($T = \tau = 90$). The S&P500 index had its largest drawdown in February 2009, marked by *lower asterisk*: it peaked in October 2007 (*lower black diamond*) and lost 70.28 % from October 2007 to February 2009. During the same period, the AORDA Portfolio 2 earned 76.39 %; therefore, the MaxDD beta of the AORDA Portfolio 2 is given by $\beta_{\rm DD} = -76.39\,\%/70.28\,\% = -1.09$

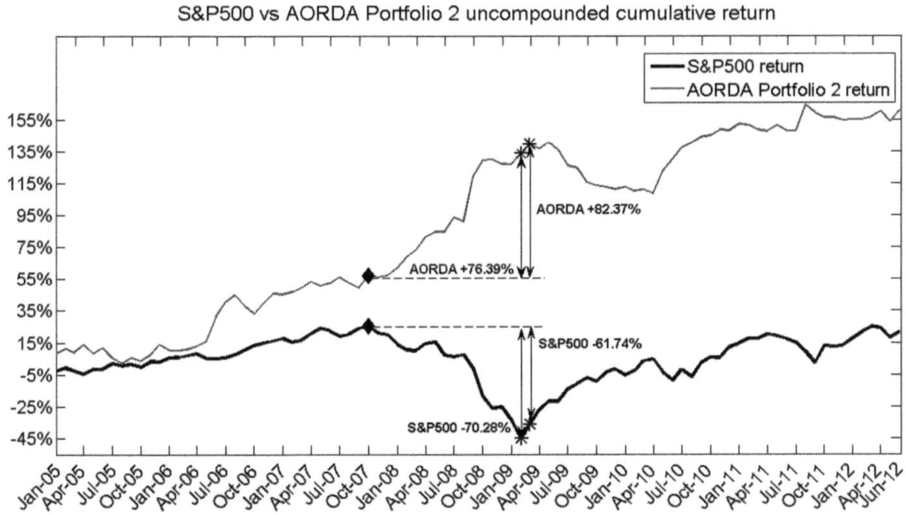

Fig. 8.4 Uncompounded monthly cumulative rates of return of the S&P500 index and of the AORDA Portfolio 2 ($T = \tau = 90$). The S&P500 index had two largest drawdowns in February 2009 (70.28 %) and in March 2009 (61.74 %) marked by *two lower asterisks*: it peaked in October 2007 (*lower black diamond*) and lost 70.28 % from October 2007 to February 2009 and 61.74 % from October 2007 to March 2009. During the same periods, the AORDA Portfolio 2 earned 76.39 % and 82.37 %, respectively. The CDaR beta of the AORDA Portfolio 2 with $\alpha = 0.978$ is given by $\beta_{\rm DD} = \frac{1}{2}(-76.39\,\% - 82.37\,\%)\,/\left(\frac{1}{2}(70.28\,\% + 61.74\,\%)\right) = -1.20$

The drawdown beta $\beta_{\rm DD}$ is illustrated with the AORDA[5] Portfolio 2 and the S&P500 index being the instrument and the "optimal" portfolio, respectively, for the period from January 2005 to June 2012 with 90 monthly periods ($T = 90$) and $\tau = 90$. Figures 8.3 and 8.4 both show the uncompounded cumulative rate of return for the AORDA Portfolio 2 and for the S&P500 index from January 2005 to June 2012. The S&P500 index peaked in October 2007, and its uncompounded cumulative rate of return from January 2005 to October 2007 was about 25.4 %. From October 2007 to February 2009, the S&P500 index lost 70.28 %, which was its maximum drawdown. During the same period, the AORDA Portfolio 2 earned 76.39 %. Consequently, the MaxDD beta ($\alpha = 1$) for AORDA Portfolio 2 is given by $\beta_{\rm DD} = -76.39\,\%/70.28\,\% = -1.09$; see Fig. 8.3. If $\alpha = 0.978$, then $(1-\alpha)\,T = 2$, and we should pick the two largest drawdowns in the history of the S&P500 index. From October 2007 to February 2009 and from October 2007 to March 2009, the S&P500 index lost 70.28 % and 61.74 %, respectively, which were the two largest drawdowns. During the same periods, the AORDA Portfolio 2 earned 76.39 % and 82.37 %, respectively. Therefore, the CDaR beta for the AORDA Portfolio 2 with $\alpha = 0.978$ is given by $\beta_{\rm DD} = \frac{1}{2}(-76.39\,\% - 82.37\,\%)\,/\left(\frac{1}{2}(70.28 + 61.74\,\%)\right) = -1.20$; see Fig. 8.4.

[5] AORDA Portfolio 2 is described at http://www.aorda.com.

Part III
Portfolio Safeguard Case Studies

Chapter 9
Portfolio Safeguard Case Studies

9.1 Case Study: Optimal Hedging of CDO Book (meanabs_dev, meanabs_pen, polynom_abs, cardn)

Background

This case study designs a portfolio of *credit default swaps* (CDS) and credit indices to hedge against changes in a *collateralized debt obligation* (CDO) book. *The hedging problem is to minimize risk of portfolio losses subject to budget and cardinality constraints* on hedge positions. The risk is measured by mean absolute deviation (MAD) and by $\mathscr{L}_1$-norm (mean absolute penalty).

CDSs are the most popular credit derivatives (Hull 2006). A CDS provides insurance against the risk of default (credit event) of a particular company. A buyer of the CDS has the right to sell bonds issued by the company for their face value when the company is in default. The buyer makes periodic payments to the seller until the end of the life of the CDS or until a default occurs. The total amount paid per year, as a percentage of the notional principal, is known as the *CDS spread*. Credit indices are indices that track CDS spreads.

A CDO is a credit derivative based on defaults of a pool of assets. Its common structure involves tranching or slicing the credit risk of the reference pool into different risk levels of increasing seniority. The losses first affect the *equity* (first loss) tranche, then the *mezzanine* tranche, and finally the *senior* and *super senior* tranches.

Reference

Hull, J. (2006): Risk Management and Financial Institutions. Prentice Hall

M. Zabarankin and S. Uryasev, *Statistical Decision Problems: Selected Concepts and Portfolio Safeguard Case Studies*, Springer Optimization and Its Applications 85,
DOI 10.1007/978-1-4614-8471-4_9,

Notation	Description
I	Number of CDS in the portfolio
$i \in \{1, \ldots, I\}$	CDS index in the portfolio
M	Number of indices in the portfolio
$m \in \{1, \ldots, M\}$	Index of the indices in the portfolio
J	Number of scenarios
$j \in \{1, \ldots, J\}$	Index of scenarios
$h = (h_1, \ldots, h_I)$	Hedge positions in CDSs (decision vector)
$\overline{h} = (\overline{h}_1, \ldots, \overline{h}_M)$	Hedge positions in indices (decision vector)
P_i	Change of the book value upon 1 basis point (bp) change in the credit spread of the ith name (the book price value of a basis point (PVBP))
$\mathbf{D} = (D_1^c, \ldots, D_I^c)$	Vector of durations of CDSs
$\overline{\mathbf{D}} = (\overline{D}_1^c, \ldots, \overline{D}_I^c)$	Vector of durations of indices
k_i	Upper bound on position in CDS i
$\overline{k}_m$	Upper bound on position in index m
BO_i	Bid-offer spread for CDS i
$\overline{BO}_m$	Bid-offer spread for index m
C	Available capital for hedging
$S_{i,j}$	Spread for CDS i on scenario j
$\overline{S}_{m,j}$	Spread for index m on scenario j
$\Delta S_{i,j} = S_{i,j} - S_{i,j-1}$	Change in the spread of CDS i on scenario j
$\Delta \overline{S}_{m,j} = \overline{S}_{m,j} - \overline{S}_{m,j-1}$	Change in the spread of index m on scenario j
Card	Upper bound on cardinality

Main Functions

- The change in portfolio value on scenario j is defined by

$$\Delta \Pi = \sum_{i=1}^{I} S_{i,j}\,(P_i + h_i D_i)\,\frac{\Delta S_{i,j}}{S_{i,j}} + \sum_{m=1}^{M} \overline{S}_{m,j}\overline{h}_m \overline{D}_m \frac{\Delta \overline{S}_{m,j}}{\overline{S}_{m,j}}.$$

- The *portfolio loss* on scenario j is defined by

$$L\left(h, \overline{h}, \Delta S_j, \Delta \overline{S}_j\right) = \sum_{i=1}^{I} S_{i,j}\,(h_i D_i - P_i)\,\frac{\Delta S_{i,j}}{S_{i,j}} + \sum_{m=1}^{M} \overline{S}_{m,j}\overline{h}_m \overline{D}_m \frac{\Delta \overline{S}_{m,j}}{\overline{S}_{m,j}}.$$

- The *MAD* function is defined by

$$\textbf{meanabs\_dev}\left(L\left(h,\overline{h},\Delta S,\Delta\overline{S}\right)\right)$$
$$=\frac{1}{J}\sum_{j=1}^{J}\left|L\left(h,\overline{h},\Delta S_j,\Delta\overline{S}_j\right)-\frac{1}{J}\sum_{i=1}^{J}L\left(h,\overline{h},\Delta S_i,\Delta\overline{S}_i\right)\right|.$$

- The *mean absolute penalty* function is defined by

$$\textbf{meanabs\_pen}\left(L\left(h,\overline{h},\Delta S,\Delta\overline{S}\right)\right)=\frac{1}{J}\sum_{j=1}^{J}\left|L\left(h,\overline{h},\Delta S_j,\Delta\overline{S}_j\right)\right|.$$

- The *polynomial absolute* function is defined by

$$\textbf{polynom\_abs}\left(h,\overline{h}\right)=\sum_{i=1}^{I}\frac{BO_i D_i}{2}|h_i|+\sum_{m=1}^{M}\frac{\overline{BO}_m\overline{D}_m}{2}\left|\overline{h}_m\right|.$$

- The *cardinality* function is defined by

$$\textbf{cardn}\left(h,\overline{h},w\right)=\sum_{i=1}^{I}u\left(\frac{BO_i D_i}{2}h_i,w\right)+\sum_{m=1}^{M}u\left(\frac{\overline{BO}_m\overline{D}_m}{2}\overline{h}_m,w\right),$$

  where

$$u(y,w)=\begin{cases}0, & \text{if } -w<y<w,\\ 1, & \text{otherwise,}\end{cases}$$

  and w is a threshold value.

9.1.1 Problem 1: problem_cdohedge_1

Problem 9.1. Minimize the hedging risk

$$\min_{h,\overline{h}}\ \textbf{meanabs\_dev}\left(L\left(h,\overline{h},\Delta S,\Delta\overline{S}\right)\right) \tag{9.1.1}$$

subject to the *budget constraint*

$$\textbf{polynom\_abs}\left(h,\overline{h}\right)\le C \tag{9.1.2}$$

and subject to *upper bounds on the positions*

$$|h_i| \le k_i, \quad i = 1, \ldots, I, \tag{9.1.3}$$

$$\left|\overline{h}_m\right| \le \overline{k}_m, \quad m = 1, \ldots, M. \tag{9.1.4}$$

Portfolio Safeguard Formulation

problem: problem_cdohedge_1, type = **minimize**
objective: objective_mad
meanabs_dev_scenarios(**matrix_scenarios**)
constraint: constraint_polinom_abs, **upper_bound = 80**, linearize = 1
polynom_abs_matrix_polynom_abs(**matrix_constraint_budget)**
box_of_variables: **lowerbounds = point_lowerbounds,**
upperbounds = point_upperbounds
solver: CAR, precision = 2, stages = 10

Problem Datasets: Dataset1

Data	http://www.ise.ufl.edu/uryasev/files/2013/03/data_problem_cdohedge_1.zip
Solution	http://www.ise.ufl.edu/uryasev/files/2011/12/solution_problem_cdohedge_1.zip
# of variables	1,004
# of scenarios	N/A
Objective value	2.65047e+06
Running time	0.89 s (PC 2.66 GHz)

9.1.2 Problem 2: problem_cdohedge_2

Problem 9.2. Minimize the hedging risk

$$\min_{h,\overline{h}} \ \textbf{meanabs\_dev}\left(L\left(h, \overline{h}, \Delta S, \Delta\overline{S}\right)\right) \tag{9.1.5}$$

subject to *the budget constraint*

$$\textbf{polynom\_abs}\left(h, \overline{h}\right) \le C, \tag{9.1.6}$$

the *constraint on cardinality*

$$\textbf{cardn}\left(h, \overline{h}, w\right) \le \text{Card}, \tag{9.1.7}$$

and subject to *upper bounds on the positions*

$$|h_i| \le k_i, \quad i = 1, \ldots, I, \tag{9.1.8}$$

$$\left|\overline{h}_m\right| \le \overline{k}_m, \quad m = 1, \ldots, M. \tag{9.1.9}$$

Portfolio Safeguard Formulation

problem: problem_cdohedge_2, type = **minimize**
objective: objective_meanabs_pen, linearize = 1
meanabs_pen_scenarios(**matrix_scenarios)**
constraint: constraint_polynom_abs, **upper_bound = 1000000**, linearize = 1
polynom_abs_matrix_polynom_abs(**matrix_constraint_budget)**
constraint: constraint_cardn_new, **upper_bound = 50**, linearize = 1
cardn_new(0.1, matrix_constraint_budget)
box_of_variables: **lowerbounds = point_lowerbounds,**
upperbounds = point_upperbounds
solver: CAR, precision = 6, stages = 10

Problem Datasets: Dataset1

Data	http://www.ise.ufl.edu/uryasev/files/2013/03/data_problem_cdohedge_2.zip
Solution	http://www.ise.ufl.edu/uryasev/files/2011/12/solution_problem_cdohedge_2.zip
# of variables	1,004
# of scenarios	50
Objective value	1.72763e+06
Running time	5.91 s (PC 2.66 GHz)

9.2 Case Study: Hedging Portfolio of Options (cvar_risk)

Background

This case study demonstrates hedging of a portfolio of options (target portfolio) by a portfolio of stocks, indices, and options (hedging portfolio). A similar case study was considered by Rockafellar and Uryasev (2000).

The composition of the target portfolio is known, whereas the hedging portfolio positions are determined by optimization (shorting is allowed). Long positions are opened at ask prices and are closed at bid prices, whereas short positions are opened at bid prices and are closed at ask prices.

The problem is to *minimize the cost of the hedging portfolio at the initial time* $t = 0$ *subject to a conditional value-at-risk (CVaR) constraint on underperformance (loss) of the hedging portfolio with respect to the target portfolio at the expiration time* $t = T$. The CVaR constraint with the confidence level α bounds the average of $(1-\alpha)\cdot 100\%$ of largest losses. Scenarios of stock and index prices at expiration are generated by bootstrapping of historical daily prices.

Reference

Rockafellar, R.T., Uryasev, S. (2000): Optimization of conditional value-at-risk. Journal of Risk **2**, 21–41

Notation	Description
T	Expiration time ($T = 62$ days)
N	Number of stocks in the hedging portfolio
$n \in \{1, \ldots, N\}$	Index of stocks
M	Number of indices in the hedging portfolio
$m \in \{1, \ldots, M\}$	Index of stock indices
S	Number of scenarios
$s \in \{1, \ldots, S\}$	Index of scenarios
$\tilde{R}^s_{nt}$	Daily return of stock n on day t under scenario s
R^s_{mt}	Daily return of index m on day t under scenario s
J_n	Number of options on stock n
$j \in \{1, \ldots, J_n\}$	Index of options on stock n
K_m	Number of options on index m
$k \in \{1, \ldots, K_m\}$	Index of options on index m
O^S_{nj}	Option j on stock n
O^I_{mk}	Option k on index m
$\tilde{K}_{nj}$	Strike price of option j on stock n
K_{mk}	Strike price of option k on index m
S^0_n	Initial price of stock n
S_n	Price of stock n at final time T
S^A_n	Ask price of stock n at initial time
S^B_n	Bid price of stock n at initial time
RL^s_n	Full return (in USD) of a long position in stock n over time T under scenario s given by $RL^s_n = S^A_n \left(\prod_{i=1}^T \tilde{R}^s_{ni} - 1\right)$

Notation	Description
RS_n^s	Full return (in USD) of a short position in stock n over time T under scenario s given by $RS_n^s = -S_n^B \left(\prod_{i=1}^T \tilde{R}_{ni}^s - 1\right)$
S_n^s	Price of stock n under scenario s at final time T given by $S_n^s = S_n^0 \prod_{i=1}^T \tilde{R}_{ni}^s$
I_m^0	Initial price of index m
I_m^s	Price of index m under scenario s at final time T given by $I_m^s = I_m^0 \prod_{i=1}^T R_{mi}^s$
$\tilde{C}_{nj}^s$	Value at expiration of option j on stock n under scenario s
C_{mk}^s	Value at expiration of option k on index m under scenario s
C_{mk}^B	Bid price of option k on index m at initial time
C_{mk}^A	Ask price of option k on index m at initial time
y_{nj}	Number of contracts on option j on stock n in the target portfolio ($y_{nj} > 0$ for a long position, $y_{nj} < 0$ for a short position)
x_{mk}^+	Number of long contracts on option O_{mk}^I in the hedging portfolio
x_n^+	Number of shares of a long position in stock n in the hedging portfolio
x_n^-	Number of shares of a short position in stock n in the hedging portfolio
x_{mk}^-	Number of short contracts on option O_{mk}^I in the hedging portfolio
$P(x)$	Initial price of the hedging portfolio
$L(x,s)$	Hedging error at final time T under scenario s
γ	Coefficient determining bound on CVaR of underperformance of the hedging portfolio
$U = 100$	Available funding in USD
$\nu = 0.15$	Diversification constant for exposures
$u_m = \dfrac{\nu U}{I_m^0}$	Upper bounds on number of contracts (long or short) on option O_{mk}^I in the hedging portfolio
$u_n = \dfrac{\nu U}{S_m^0}$	Upper bounds on number of shares of a long (short) position in stock n in the hedging portfolio

Main Functions

- The value of option j on stock n at expiration under scenario s is given by

$$\tilde{C}_{nj}^s = \begin{cases} \max\left\{0, S_n^s - \tilde{K}_{nj}\right\} & \text{for call option,} \\ \max\left\{0, \tilde{K}_{nj} - S_n^s\right\} & \text{for put option.} \end{cases}$$

- The value of option k on index m at expiration under scenario s is given by

$$C^s_{mk} = \begin{cases} \max\{0, I^s_m - K_{mk}\} & \text{for call option,} \\ \max\{0, K_{mk} - I^s_m\} & \text{for put option.} \end{cases}$$

- The *initial price of the hedging portfolio* is given by

$$P(x) = \sum_{m=1}^{M} \sum_{k=1}^{K_m} (x^+_{mk} C^A_{mk} - x^-_{mk} C^B_{mk}) + \sum_{n=1}^{N} (x^+_n S^A_n - x^-_n S^B_n).$$

- The *hedging error* at the final time T under scenario s is defined by

$$L(x,s) = \sum_{n=1}^{N} \sum_{j=1}^{J_n} y_{nj}\, \tilde{C}^s_{nj} - \left[\sum_{m=1}^{M} \sum_{k=1}^{K_m} \left(x^+_{mk} - x^-_{mk}\right) C^s_{mk} + \sum_{n=1}^{N} (x^+_n RL^s_n + x^-_n RS^s_n) \right],$$

  where the first term and the second term (in square brackets) are the values of the target portfolio and the hedging portfolio, respectively, at the final time T.
- CVaR of the hedging error $L(x,s)$ is *CVaR for loss* [CVaR^-_α, see (8.6.1)] determined by

$$\textbf{cvar\_risk}_\alpha(L(x,s)) = \min_{c \in \mathbb{R}} \left(c + \frac{1}{1-\alpha} \sum_{j=1}^{J} p_j \max\left\{L_j(x,s) - c, 0\right\} \right).$$

Problem: problem_cs_hedging_portfolio_of_options

Problem 9.3. Minimize the initial portfolio value

$$\min_x P(x) \tag{9.2.1}$$

subject to a *CVaR constraint on the hedging error*

$$\textbf{cvar\_risk}_\alpha(L(x,s)) \le \gamma\, U \tag{9.2.2}$$

and subject to *box constraints on the positions*

$$0 \le x^+_{mk} \le u_m, \quad 0 \le x^-_{mk} \le u_m, \qquad k = 1, \ldots, K_m, \quad m = 1, \ldots, M, \tag{9.2.3}$$

$$0 \le x^+_n \le u_n, \quad 0 \le x^-_n \le u_n, \qquad n = 1, \ldots, N. \tag{9.2.4}$$

The optimization problem (9.2.1)–(9.2.4) is solved for Dataset1 with $\gamma = 5.30$ and $\nu = 0.15$ and for Dataset2 with $\gamma = 5.25$ and $\nu = 0.15$.

Portfolio Safeguard Formulation

problem: problem_cs_hedging_portfolio_of_options_5p3, type = **minimize**
objective: objective_Options
linear_initial_price(**matrix_options_prices**)
constraint: constraint_CVaR_Options, **upper_bound = 530**, linearize = 1
cvar_risk_hedging_constraint(**0.94999999, matrix_options_scenarios_45**)
box_of_variables: **lowerbounds = point_lowerbounds,**
upperbounds = point_upperbounds
solver: VAN, precision = 6

Problem Datasets: Dataset1

Data	http://www.ise.ufl.edu/uryasev/files/2013/04/data_problem_cs_hedging_portfolio_of_options_long.zip
Solution	http://www.ise.ufl.edu/uryasev/files/2011/12/solution_problem_cs_hedging_portfolio_of_options_long.zip
# of variables	18
# of scenarios	211,618
Objective value	0.182599
Running time	52.74 s (PC 2.66 GHz)

Problem Datasets: Dataset2

Data	http://www.ise.ufl.edu/uryasev/files/2013/04/data_problem_cs_hedging_portfolio_of_options_short.zip
Solution	http://www.ise.ufl.edu/uryasev/files/2011/12/solution_problem_cs_hedging_portfolio_of_options_short.zip
# of variables	30
# of scenarios	12,925
Objective value	0.240832
Running time	0.65 s (PC 2.66 GHz)

9.3 Case Study: Mortgage Pipeline Hedging (st_dev, meanabs_dev, cvar_dev, var_risk, var_risk_g)

Background

Usually, mortgage lenders sell mortgages in the secondary market. Alternatively, they can exchange mortgages for mortgage-backed securities (MBSs) and then sell MBSs in the secondary market (Lederman 1997). This case study demonstrates hedging against the risk in the mortgage underwriting process known as "pipeline." Mortgage lenders commit to a mortgage interest rate while the loan is in process, typically for a period of 30–60 days. If the rate rises before the loan goes to closing, the value of the loan declines and the lender sells the loan at a lower price. The risk that mortgages in process will fall in value prior to their sale is known as *mortgage pipeline risk*. Lenders often hedge this exposure either by selling forward their expected closing volume or by shorting either US Treasury notes or futures contracts (Taglia 2003).

Fallout refers to the percentage of loan commitments that do not go to closing (Hakim et al. 1999). It affects the mortgage pipeline risk. As interest rates fall, the fallout rises because borrowers locked in a mortgage rate are more likely to find better rates with another lender. Conversely, as rates rise, the percentage of loans that close increases. So, the fallout alters the size of the pipeline position to be hedged against and, as a result, affects the required size of the hedging instrument: at lower rates, fewer rate loans will close and a smaller position in the hedging instrument is needed. To hedge against the fallout risk, lenders often use options on US Treasury note futures (Cusatis and Thomas 2005).

This case study compares five *hedging strategies*, each of which solves the following *pipeline risk hedging problem*: *minimize a deviation measure of the underperformance of a hedging portfolio with respect to a hedging target.* Three hedging instruments are available: 5% MBS forward, 5.5% MBS forward, and call options on 10-year Treasury note futures. Short sales are allowed, and transaction costs are ignored. The strategies differ only in the deviation measures they use: *standard deviation*, *MAD*, *CVaR deviation*, *two-tailed 75%-VaR deviation*[1], and *two-tailed 90%-VaR deviation*.

Each hedging strategy undergoes the following out-of-sample testing. Available 1,000 scenarios are split into ten equal-size groups. Hedging portfolio positions are found based on nine groups of 900 scenarios, and portfolio losses are evaluated for the remaining group of 100 scenarios (out-of-sample group). This procedure is repeated ten times, each time with a different out-of-sample group, and then all out-of-sample losses, aggregated from ten runs (1,000 scenarios), are measured by the corresponding deviation measure.

---

[1]Two-tailed α-VaR deviation for loss $L(\theta, X)$ is defined as $\mathrm{VaR}_\alpha(L(\theta, X)) + \mathrm{VaR}_\alpha(-L(\theta, X))$, i.e., $\alpha \cdot 100\%$ percentile of the loss distribution plus $\alpha \cdot 100\%$ percentile of the profit distribution.

In addition, three downside risk measures, *90%-CVaR*, *90%-VaR*, and *max loss*, are calculated on the aggregated out-of-sample losses corresponding to the five strategies. All three risk measures have the smallest values for the strategy with two-tailed 90%-VaR deviation, whereas all of them have the largest values for the strategy with standard deviation. The three risk measures show good results for the strategy with CVaR deviation.

References

Cusatis, P.J., Thomas, M.R. (2005): Hedging Instruments and Risk Management. McGraw-Hill

Taglia, P. (2003): Risk management case study: how mortgage lenders use futures to hedge pipeline risk. Futures Industry Magazine, September/October

Lederman, J. (1997): Handbook of Secondary Marketing. Mortgage Bankers Association of America

Hakim, S., Rashidian, M., Rosenblatt, E. (1999): Measuring the fallout risk in the mortgage pipeline. The Journal of Fixed Income **9**, 62–75

Notation	Description
I	Number of hedging instruments
$i \in \{1, \ldots, I\}$	Index of hedging instrument in the portfolio
J	Number of scenarios
$j \in \{1, \ldots, J\}$	Scenario index
$p_j = 1/J$	Probability of scenario j; $j = 1, \ldots, J$
$x = (x_1, \ldots, x_I)$	Decision vector (positions in hedging instruments)
θ_0	Return of the benchmark (target)
θ_{0j}	Return of the benchmark under scenario j
θ_{ij}	Return of instrument i under scenario j $i = 1, \ldots, I$, $j = 1, \ldots, J$
$\theta = (\theta_0, \theta_1, \ldots, \theta_I)$	Random vector of returns of the benchmark and of hedging instruments
$\theta = (\theta_{0j}, \theta_{1j}, \ldots, \theta_{Ij})$	Vector of returns of instruments under scenario j
$L(x, \theta) = \theta_0 - \sum_{i=1}^{I} \theta_i x_i$	Loss function
$L(x, \theta_j) = \theta_{0j} - \sum_{i=1}^{I} \theta_{ij} x_i$	Loss function under scenario j

Main Functions

- Standard deviation of the loss $L(x, \theta)$ is defined by

$$\textbf{st\_dev}(L(x, \theta)) = \sqrt{\sum_{j=1}^{J} p_j \left(L(x, \theta_j) - E[L(x, \theta)]\right)^2},$$

where

$$E[L(x,\theta)] = \sum_{j=1}^{J} p_j L(x,\theta_j).$$

- MAD of the loss $L(x,\theta)$ is implemented by the function

$$\mathbf{meanabs\_dev}(L(x,\theta)) = \sum_{j=1}^{J} p_j \left| L(x,\theta_j) - E[L(x,\theta)] \right|.$$

- *CVaR deviation* of the loss $L(x,\theta)$ is implemented by the function

$$\begin{aligned}\mathbf{cvar\_dev}_\alpha(L(x,\theta)) &= \mathbf{cvar\_risk}_\alpha\left(L(x,\theta) - E[L(x,\theta)]\right),\\ &\equiv \mathbf{cvar\_risk}_\alpha\left(L(x,\theta)\right) + E[L(x,\theta)],\end{aligned}$$

where $\mathbf{cvar\_risk}_\alpha$ is *CVaR for loss* [CVaR_α^-, see (8.6.1)] defined by

$$\mathbf{cvar\_risk}_\alpha(L(x,\theta)) = \min_{c\in\mathbb{R}}\left(c + \frac{1}{1-\alpha}\sum_{j=1}^{J} p_j \max\left\{L_j(x,\theta) - c, 0\right\}\right).$$

- Two-tailed α-VaR deviation of the loss $L(x,\theta)$ is defined by

$$\mathrm{TwoTailVar}_\alpha^\Delta\left(L(x,\theta)\right) = \mathrm{VaR}_\alpha\left(L(x,\theta)\right) + \mathrm{VaR}_\alpha\left(-L(x,\theta)\right)$$

and is implemented as

$$\mathrm{TwoTailVar}_\alpha^\Delta\left(L(x,\theta)\right) = \mathbf{var\_risk}_\alpha(L(x,\theta)) + \mathbf{var\_risk\_g}_\alpha(L(x,\theta)),$$

where

$$\mathbf{var\_risk}_\alpha(L(x,\theta)) = \arg\min_{c\in\mathbb{R}}\left(c + \frac{1}{1-\alpha}\sum_{j=1}^{J} p_j \max\left\{L_j(x,\theta) - c, 0\right\}\right)$$

and

$$\mathbf{var\_risk\_g}_\alpha(L(x,\theta)) = \mathbf{var\_risk}_\alpha(-L(x,\theta)).$$

9.3.1 Problem 1: problem_cvar_dev

Problem 9.4. Minimize CVaR deviation of the loss $L(x,\theta)$ at $\alpha = 0.9$:

$$\min_x \mathbf{cvar\_dev}_\alpha\left(L(x,\theta)\right). \tag{9.3.1}$$

Portfolio Safeguard Formulation

problem: problem_CVaRDev, type = **minimize**
objective: objective_CVaRDev90
cvar_dev_scenarios(**0.9, matrix_scenarios_5**)
box_of_variables:
solver: VAN, precision = 6

Problem Datasets: Dataset1

Data	http://www.ise.ufl.edu/uryasev/files/2013/03/data_problem_cvar_dev.zip
Solution	http://www.ise.ufl.edu/uryasev/files/2011/12/solution_problem_cvar_dev.zip
# of variables	3
# of scenarios	900
Objective value	0.469855
Running time	0.03 s (PC 2.66 GHz)

9.3.2 Problem 2: problem_meanabs_dev

Problem 9.5. Minimize MAD of the loss $L(x, \theta)$:

$$\min_x \textbf{meanabs\_dev}(L(x, \theta)). \tag{9.3.2}$$

Portfolio Safeguard Formulation

problem: problem_MeanAbsDev, type = **minimize**
objective: objective_MeanAbsDev, linearize = 1
meanabs_dev_scenarios(**matrix_scenarios_1**)
box_of_variables:
solver: VAN, precision = 6

Problem Datasets: Dataset1

Data	http://www.ise.ufl.edu/uryasev/files/2013/03/data_problem_meanabs_dev.zip
Solution	http://www.ise.ufl.edu/uryasev/files/2011/12/solution_problem_meanabs_dev.zip
# of variables	3
# of scenarios	900
Objective value	0.434015
Running time	0.01 s (PC 2.66 GHz)

9.3.3 Problem 3: problem_std_dev

Problem 9.6. Minimize standard deviation of the loss $L(x, \theta)$:

$$\min_x \mathbf{st\_dev}\,(L(x, \theta))\,. \tag{9.3.3}$$

Portfolio Safeguard Formulation

problem: problem_StdDev, type = **minimize**
objective: objective_StdDev, linearize = 1
st_dev_scenarios(**matrix_scenarios_1)**
box_of_variables:
solver: VAN, precision = 6

Problem Datasets: Dataset1

Data	http://www.ise.ufl.edu/uryasev/files/2013/03/data_problem_std_dev.zip
Solution	http://www.ise.ufl.edu/uryasev/files/2011/12/solution_problem_std_dev.zip
# of variables	3
# of scenarios	900
Objective value	0.996590
Running time	0.01 s (PC 2.66 GHz)

9.3.4 Problem 4: problem_two_tailed_var75

Problem 9.7. Minimize two-tailed 75%-VaR deviation of the loss $L(x,\theta)$:

$$\min_x \text{TwoTailVar}^{\Delta}_{0.75}\left(L(x,\theta)\right). \qquad (9.3.4)$$

Portfolio Safeguard Formulation

problem: problem_TwoTailedVaR75, type = **minimize**
objective: objective_TwoTailedVaR75, linearize = 1
var_risk_g_scenarios_75(**0.75, matrix_scenarios_1**)
var_risk_scenarios_75(**0.75, matrix_scenarios_1**)
box_of_variables:
solver: VAN, precision = 6, stages = 30

Problem Datasets: Dataset1

Data	http://www.ise.ufl.edu/uryasev/files/2013/03/data_problem_two_tailed_var75.zip
Solution	http://www.ise.ufl.edu/uryasev/files/2011/12/solution_problem_two_tailed_var75.zip
# of variables	3
# of scenarios	900
Objective value	0.030025
Running time	2.08 s (PC 2.66 GHz)

9.3.5 Problem 5: problem_two_tailed_var90

Problem 9.8. Minimize two-tailed 90%-VaR deviation of the loss $L(x,\theta)$:

$$\min_x \text{TwoTailVar}^{\Delta}_{0.90}\left(L(x,\theta)\right). \qquad (9.3.5)$$

Portfolio Safeguard Formulation

problem: problem_TwoTailedVaR90, type = **minimize**
objective: objective_TwoTailedVaR90, linearize = 1
var_risk_g_scenarios_90(**0.9, matrix_scenarios_1**)
var_risk_scenarios_90(**0.9, matrix_scenarios_1**)
box_of_variables:
solver: VAN, precision = 6, stages = 30

Problem Datasets: Dataset1

Data	http://www.ise.ufl.edu/uryasev/files/2013/03/data_problem_two_tailed_var90.zip
Solution	http://www.ise.ufl.edu/uryasev/files/2011/12/solution_problem_two_tailed_var90.zip
# of variables	3
# of scenarios	900
Objective value	0.256648
Running time	1.6 s (PC 2.66 GHz)

9.4 Case Study: Cash Matching Bond Portfolio (max_dev, avg)

Background

This case study *designs a bond portfolio with minimal initial cost subject to the constraint that payments made from the portfolio match cash obligations over given time period* (*cash matching* optimization problem). It is assumed that bonds in the portfolio are of various maturities and pay coupons as well as face values at different time periods. A similar problem is discussed in Luenberger (1998), p. 108.

Reference

Luenberger, D.G. (1998): Investment Science. Oxford University Press

Notation	Description
J	Number of scenarios (time periods)
$j \in \{1, \ldots, J\}$	Index of scenarios (years)
I	Number of bonds in the replicating portfolio
$i \in \{1, \ldots, I\}$	Bond index in the replicating portfolio
$p_j = 1/J$	Probability of scenario j; $j = 1, \ldots, J$
x_i	Number of shares of bond i in the portfolio
$x = (x_1, \ldots, x_I)$	Decision vector
a_i	Price of bond i at initial time
b_j	Liability at the end of year j (in scenario j)
c_{ij}	Amount of receipts at the end of year j (corresponding to scenario j) resulting from one share of bond i
r	Annual risk-free rate (constant over time periods $j = 1, \ldots, J$)
$\theta_{j0} = \dfrac{b_j}{(1+r)^j}$	Liability b_j discounted to present time
$\theta_{ji} = \dfrac{c_{ij}}{(1+r)^j}$	Bond payment c_{ij}, $i = 1, \ldots, I$, discounted to present time
θ_0	Random variable with possible realizations $(\theta_{10}, \ldots, \theta_{J0})$
θ_i	Random variable with possible realizations $(\theta_{1i}, \ldots, \theta_{Ji})$, $i = 1, \ldots, I$
$\theta = (\theta_0, \theta_1, \ldots, \theta_I)$	Random scenario vector
w	Bound on the underperformance of the replicating portfolio versus the liability
K	Penalty coefficient
$L(x, \theta) = \theta_0 - \sum_{i=1}^{I} \theta_i x_i$	Loss function, which is the underperformance of the replicating portfolio discounted to the initial time versus the liability at the end of year j (in scenario j)
$L_j(x) = \theta_{0j} - \sum_{i=1}^{I} \theta_{ij} x_i$	Loss function under scenario j

Main Functions

- The *average loss* is the expected value of $L(x, \theta)$:

$$\mathbf{avg}(L(x,\theta)) = E[L(x,\theta)] = \sum_{j=1}^{J} p_j L_j(x).$$

- The *maximum deviation* for the loss $L(x, \theta)$ is implemented by the function

$$\mathbf{max\_dev}(L(x,\theta)) = \max_{1 \le j \le J} L_j(x) - E[L(x,\theta)].$$

Problem: problem_cs_cash_matching_60

Problem 9.9. Minimize the portfolio cost

$$\min_x \sum_{i=1}^{I} a_i x_i \tag{9.4.1}$$

subject to the mean self-financing constraint

$$\mathbf{avg}(L(x,\theta)) = 0 \tag{9.4.2}$$

and the cash matching constraint

$$\mathbf{max\_dev}(L(x,\theta)) \le w \tag{9.4.3}$$

and subject to no-shorting constraints on the positions

$$x_i \ge 0, \qquad i = 1, \ldots, I. \tag{9.4.4}$$

The constraint (9.4.3) may render the problem infeasible. To guarantee feasibility, (9.4.3) can include a nonnegative auxiliary variable z, and then z is added to the objective function (9.4.1) with a nonnegative penalty coefficient K. The following formulation illustrates this approach.

Problem 9.10. Minimize the portfolio cost with a penalty function

$$\min_{x,z} \left(\sum_{i=1}^{I} a_i x_i + Kz \right) \tag{9.4.5}$$

subject to the mean self-financing constraint

$$\mathbf{avg}(L(x,\theta)) = 0 \tag{9.4.6}$$

and the cash matching constraint

$$\mathbf{max\_dev}(L(x,\theta)) - z \le w \tag{9.4.7}$$

and subject to no-shorting constraints on the positions and subject to nonnegativity of the auxiliary variable

$$x_i \geq 0, \qquad i = 1, \ldots, I, \qquad z \geq 0. \tag{9.4.8}$$

Portfolio Safeguard Formulation

problem: problem_CS_Cash_Matching_60, type = **minimize**
objective: objective_linrar_plus_penalty, linearize = 0
linear_portfolio_cost(**matrix_cs_cash_matching_max_dev_avg_objective)**
+100·**variable_xa(xa)**
constraint: constraint_zero_AVG, lower_bound = 0, upper_bound = 0
avg_CS_underperformance(**matrix_cs_cash_matching_max_dev_avg_loss)**
constraint: constraint_MAX_DEV_minus_penalty, **upper_bound = 60**
max_dev_underperformance(**matrix_cs_cash_matching_max_dev_avg_loss)**
−**variable_xa(xa)**
box_of_variables: **lowerbounds = point_lowerbounds**
solver: VAN, precision = 6

Problem Datasets: Dataset1

Data	http://www.ise.ufl.edu/uryasev/files/2013/04/data_problem_cs_cash_matching_60.zip
Solution	http://www.ise.ufl.edu/uryasev/files/2011/12/solution_problem_cs_cash_matching_60.zip
# of variables	11
# of scenarios	6
Objective value	2.29723e+03
Running time	0.01 s (PC 2.66 GHz)

9.5 Case Study: Cash Flow Matching with CVaR Constraints (linear, cvar_max_risk)

Background

A *cash flow matching problem* optimizes a portfolio of given financial instruments (typically bonds) to match existing cash obligations (liabilities) over several time periods in the form of portfolio payments (dividends). Typically, the problem is formulated in a stochastic programming setting; see Hiller and Eckstein (1993), Zenios (1995), and Consigli and Dempster (1998).

Iyengar and Ma (2009) *minimized the initial cost of a bond portfolio subject to a CVaR constraint on matching liabilities/obligations over several time periods* with the assumptions:

1. The time horizon of the liabilities is longer than the maturities of available bonds.
2. Bonds of various maturities pay coupons and face values at different time periods.
3. Bond purchase decisions are made each period to generate cash flow for meeting future obligations.
4. Bond interest rates are uncertain.
5. Bond future prices are uncertain.

This case study solves Iyengar and Ma's cash flow matching problem with a set of scenarios for bond prices provided by Prof. Ken Kortanek (for benchmarking purposes) and generated with Hull and White (1990) methodology; see Shang and Uryasev (2011) for details on the problem formulation and on numerical results.

References

Hiller, R., Eckstein, J. (1993): Stochastic dedication: designing fixed income portfolios using massively parallel Benders decomposition. Management Science **39**(11), 1422–1438

Zenios, S. (1995): Asset/Liability management under uncertainty for fixed-income securities. Annals of Operations Research **59**(1), 77–97

Consigli, G., Dempster, M. (1998): Dynamic stochastic programming for asset-liability management. Annals of Operations Research **81**, 131–161

Iyengar, G., Ma, A. (2009): Cash flow matching: a risk management approach. North American Actuarial Journal **13**(3), 370–378

Hull, J., White, A. (1990): Pricing interest rate derivative securities. Review of Financial Studies **3**(4), 573–592

Shang, D., Uryasev, S. (2011): Cash flow matching problem with CVaR. Research Report #2011-1. http://www.ise.ufl.edu/uryasev/files/2011/11/CashFlow_WP.pdf

Notation	Description
T	Number of time periods
$t \in \{0, \dots, T\}$	Index of time periods
J	Number of bonds available for investment
$j \in \{1, \dots, J\}$	Bond index
p_{0j}	Initial price of bond j
$p_0 = (p_{01}, \dots, p_{0J})$	Vector of bond initial prices
p_{tj}	Random price of bond j at the end of time period t; $t = 1, \dots, T$
$p_t = (p_{t1}, \dots, p_{tJ})$	Random vector of bond prices at the end of time period t
K	Number of scenarios for bond prices
$k \in \{1, \dots, K\}$	Scenario index
π_k	Probability of scenario k
c_{stj}	Cash flow after time period t from bond j purchased at time period $s < t$; $j = 1, \dots, J$, $t = 1, \dots, T$, $s = 0, \dots, t-1$
$c_{st} = (c_{st1}, \dots, c_{stJ})$	Vector of cash flows after time step t from all bonds purchased at time period s; $t = 1, \dots, T$, $s = 0, \dots, t-1$
x_{0j}	Number of shares of initially purchased bond j; $j = 1, \dots, J$
$x_0 = (x_{01}, \dots, x_{0J})$	Vector of initially purchased bonds
x_{tj}	Number of shares of bond j purchased at time period t; $j = 1, \dots, J$, $t = 1, \dots, T$
$x_t = (x_{t1}, \dots, x_{tJ})$	Vector of numbers of shares of bonds purchased at time period t; $t = 1, \dots, T$
$u_t = (u_{t1}, \dots, u_{tJ})$	Vector of upper bounds for x_t; $t = 0, \dots, T$
l_t	Liability at time period t; $t = 1, \dots, T$

Main Functions

- The random loss function at time period t is defined by

$$L_t(x_0, \dots, x_t) = l_t + p_t \cdot x_t - \sum_{s=0}^{t-1} c_{st} \cdot x_s, \qquad t = 1, \dots, T.$$

- **cvar_max_risk**$_\alpha$ $(L_1(x_0, x_1), L_2(x_0, x_1, x_2), \dots, L_T(x_0, \dots, x_T))$ is the *CVaR max risk for loss function* defined by

$$\mathbf{cvar\_max\_risk}_\alpha\,(L_1(x_0,x_1), L_2(x_0,x_1,x_2),\ldots,L_T(x_0,\ldots,x_T))$$
$$= \mathbf{cvar\_risk}_\alpha\left(\max_{t=1,\ldots,T} L_t(x_0,\ldots,x_t)\right),$$

where $\mathbf{cvar\_risk}_\alpha(Y)$ is *CVaR for loss* [CVaR_α^-, see (8.6.1)]:

$$\mathbf{cvar\_risk}_\alpha(Y) = \min_{c\in\mathbb{R}}\left(c + \frac{1}{1-\alpha}\sum_{k=1}^{K}\pi_k \max\{y_k - c, 0\}\right)$$

with Y being a discretely distributed random variable (loss) that assumes values $y_1,\ldots,y_K$ with probabilities $\pi_1,\ldots,\pi_K$ corresponding to K bond price scenarios.

Problem Formulation

Problem 9.11 (Cash flow matching problem with CVaR constraint). Minimize the cost of portfolio

$$\min_{x_0,\ldots,x_T} p_0^\top x_0 \tag{9.5.1}$$

subject to the downside risk constraint

$$\mathbf{cvar\_max\_risk}_\alpha\,(L_1(x_0,x_1), L_2(x_0,x_1,x_2),\ldots,L_T(x_0,\ldots,x_T)) \le \text{const} \tag{9.5.2}$$

and subject to bounds on the decision variables

$$0 \le x_t \le u_t, \qquad t = 0,\ldots,T. \tag{9.5.3}$$

The following four PSG formulations of the problem (9.5.1)–(9.5.3) differ only in the datasets that they use, namely, Dataset1, Dataset2, Dataset3, and Dataset4. Dataset1 has a purely demonstrative purpose, whereas Dataset2, Dataset3, and Dataset4 are of large size and from real-life problems.

Portfolio Safeguard Formulation

problem: problem_Cash_flow_matching_Big1_0_0.2, type = **minimize**
objective: objective_cvar_max_risk
linear_1(matrix_0)
constraint: constraint_cvar_max, **upper_bound = 0**
cvar_max_risk_1(0.2, matrix_1, matrix_2, . . . , matrix_120)
box_of_variables: **lowerbounds = 0**, **upperbounds = 10**
solver: VAN, precision = 6, stages = 6

Problem Datasets: Dataset1

Data	http://www.ise.ufl.edu/uryasev/files/2013/04/data_problem_cash_flow_matching_big1.zip
Solution	http://www.ise.ufl.edu/uryasev/files/2011/12/solution_problem_cash_flow_matching_big1_0_0.2.zip
# of variables	1,331
# of scenarios	200
Objective value	1.16792e+03
Running time	175.9 s (PC 3.40 GHz)

Portfolio Safeguard Formulation

problem: problem_Cash_flow_matching_Big1_2_0.9, type = **minimize**
objective: objective_cvar_max_risk
linear_1(matrix_0)
constraint: constraint_cvar_max, **upper_bound = 2**
cvar_max_risk_1(0.9, matrix_1, matrix_2, . . . , matrix_120)
box_of_variables: **lowerbounds = 0**, **upperbounds = 10**
solver: VAN, precision = 6, stages = 6

Problem Datasets: Dataset2

Data	http://www.ise.ufl.edu/uryasev/files/2013/04/data_problem_cash_flow_matching_big1.zip
Solution	http://www.ise.ufl.edu/uryasev/files/2011/12/solution_problem_Cash_flow_matching_Big1_2_0.9.zip
# of variables	1,331
# of scenarios	200
Objective value	1.12369e+03
Running time	233.3 s (PC 3.40 GHz)

Portfolio Safeguard Formulation

problem: problem_Cash_flow_matching_Big2_0_0.2, type = **minimize**
objective: objective_cvar_max_risk
linear_1(matrix_0)
constraint: constraint_cvar_max, **upper_bound = 0**
cvar_max_risk_1(0.2, matrix_1, matrix_2, ... , matrix_120)
box_of_variables: **lowerbounds = 0**, **upperbounds = 10**
solver: VAN, precision = 6, stages = 6

Problem Datasets: Dataset3

Data	http://www.ise.ufl.edu/uryasev/files/2013/04/data_problem_cash_flow_matching_big2.zip
Solution	http://www.ise.ufl.edu/uryasev/files/2011/12/solution_problem_Cash_flow_matching_Big2_0_0.2.zip
# of variables	1,331
# of scenarios	200
Objective value	1.16792e+03
Running time	122.0 s (PC 3.40 GHz)

Portfolio Safeguard Formulation

problem: problem_Cash_flow_matching_Big2_2_0.9, type = **minimize**
objective: objective_cvar_max_risk
linear_1(matrix_0)
constraint: constraint_cvar_max, **upper_bound = 2**
cvar_max_risk_1(0.9, matrix_1, matrix_2, ... , matrix_120)
box_of_variables: **lowerbounds = 0**, **upperbounds = 10**
solver: VAN, precision = 6, stages = 6

Problem Datasets: Dataset4

Data	http://www.ise.ufl.edu/uryasev/files/2013/04/data_problem_cash_flow_matching_big2.zip
Solution	http://www.ise.ufl.edu/uryasev/files/2011/12/solution_problem_Cash_flow_matching_Big2_2_0.9.zip
# of variables	1,331
# of scenarios	200
Objective value	1.12369e+03
Running time	145.4 s (PC 3.40 GHz)

9.6 Case Study: Relative Entropy Minimization (entropyr)

Background

The *relative entropy*, also known as *Kullback–Leibler divergence*, measures closeness of two probability distributions and is implemented in PSG by the function *entropyr*. The principle of *minimum relative entropy* aims to find a probability distribution p as close to some prior probability distribution q as possible subject to available information about p. This case study minimizes the relative entropy subject to linear constraints on p.

Notation	Description
I	Number of atoms in the distribution
$q = (q_1, \dots, q_I)$	Vector of prior probabilities
$p = (p_1, \dots, p_I)$	Decision vector (unknown probabilities)
A	Matrix in the linear equality constraint
b	Vector in the right-hand side of the linear equality constraint
l_i	Lower bound on decision variable p_i
u_i	Upper bound on decision variable p_i

Main Functions

Relative entropy (1.5.3) is implemented by the PSG function

$$\textbf{entropyr}(p) = \sum_{i=1}^{I} p_i \ln \frac{p_i}{q_i} \qquad \text{for } p_i \geq 0, \quad q_i > 0, \quad i = 1, \dots, I.$$

Problem: problem_entropyr

Problem 9.12. Minimize relative entropy

$$\min_{p} \mathbf{entropyr}(p) \tag{9.6.1}$$

subject to the constraint on the sum of decision variables (probabilities)

$$\sum_{i=1}^{I} p_i = 1, \tag{9.6.2}$$

subject to linear equality constraints

$$Ap = b, \tag{9.6.3}$$

and subject to box constraints

$$l_i \leq p_i \leq u_i, \qquad i = 1, \dots, I. \tag{9.6.4}$$

Portfolio Safeguard Formulation

problem: problem_entropyr_250, type = **minimize**
objective: objective_entropyr, linearize = 1
entropyr_1(**matrix_p**)
constraint: constraint_1, **lower_bound = vector_beq**, **upper_bound = vector_beq**
linearmulti_matrix_Aeq(**matrix_aeq**)
box_of_variables: **lowerbounds = point_lowerbounds**
solver: BULDOZER, precision = 6, stages = 6

Problem Datasets: Dataset1

Data	http://www.ise.ufl.edu/uryasev/files/2013/03/data_problem_entropyr_250.zip
Solution	http://www.ise.ufl.edu/uryasev/files/2011/12/solution_problem_entropyr_250.zip
# of variables	250
Objective value	2.16727
Running time	0.01 s (PC 2.66 GHz)

Portfolio Safeguard Formulation

problem: problem_entropyr_100th, type = **minimize**
objective: objective_entropyr, linearize = 1
entropyr_h(**matrix_h**)
constraint: constraint_aeq, **lower_bound = vector_beq**,
upper_bound = vector_beq
linearmulti_aeq(**matrix_aeq**)
box_of_variables: **lowerbounds = 1e**−13
solver: BULDOZER, precision = 9, stages = 30

Problem Datasets: Dataset2

Data	http://www.ise.ufl.edu/uryasev/files/2013/03/data_problem_entropyr_100th.zip
Solution	http://www.ise.ufl.edu/uryasev/files/2011/12/solution_problem_entropyr_100th.zip
# of variables	100,000
Objective value	1.92307
Running time	4.82 s (PC 2.66 GHz)

9.7 Case Study: Portfolio Replication with Risk Constraints (meanabs_pen, cvar_risk)

Background

This case study replicates the S&P100 index by a portfolio of 30 specified stocks from the S&P100 index. A *portfolio replication* problem *minimizes the mean absolute penalty* ($\mathscr{L}_1$*-norm*) *of the replication error subject to a constraint that CVaR of the error does not exceed a specified threshold.* The CVaR constraint controls underperformance of the portfolio with respect to the S&P100 index. The following 30 stocks are chosen (tickers)—GD, UIS, NSM, ORCL, CSCO, HET, BS, TXN, HM, INTC, RAL, NT, MER, KM, BHI, GEN, HAL, BDK, HWP, LTD, BAC, AVP, AXP, AA, BA, AGC, BAX, AIG, AN, and AEP—and scenario matrices are generated based on historical data of stock prices.

Details of this case study can be found in Rockafellar and Uryasev (2002). For other references on portfolio replication, see, for instance, Andrews et al. (1986), Beasley et al. (2003), Buckley and Korn (1998), Connor and Leland (1995), Dahl et al. (1993), Konno and Wijayanayake (2001), Rudd (1980), and Toy and Zurack (1989).

References

Andrews, C., Ford, D., Mallinson, K. (1986): The design of index funds and alternative methods of replication. The Investment Analyst **82**, 16–23

Beasley, J.E., Meade, N., Chang, T.-J. (2003): An evolutionary heuristic for the index tracking problem. European Journal of Operational Research **148**(3), 621–643

Buckley, I.R.C., Korn, R. (1998): Optimal index tracking under transaction costs and impulse control. International Journal of Theoretical and Applied Finance **1**(3), 315–330

Connor, G., Leland, H. (1995): Cash management for index tracking. Financial Analysts Journal **51**(6), 75–80

Dahl, H., Meeraus, A., Zenios, S.A. (1993): Some financial optimization models: I Risk management. In: Zeniosh, S.A. (Ed.), Financial Optimization. Cambridge University Press, Cambridge, 3–36

Konno, H., Wijayanayake, A. (2001): Minimal cost index tracking under nonlinear transaction costs and minimal transaction unit constraints. International Journal of Theoretical and Applied Finance **4**(6), 939–957

Rockafellar, R.T. and Uryasev, S. (2002): Conditional value-at-risk for general loss distributions. Journal of Banking and Finance **26**(7), 1443–1471

Rudd, A. (1980): Optimal selection of passive portfolios. Financial Management **9**(1), 57–66

Toy, W.M., Zurack, M.A. (1989): Tracking the Euro-Pac index. The Journal of Portfolio Management **15**(2), 55–58

Notation	Description
I	Number of instruments in the portfolio replicating the S&P100 index
$i \in \{1, \ldots, I\}$	Index of instruments in the replicating portfolio
J	Number of scenarios (days)
$j \in \{1, \ldots, J\}$	Index of scenarios
U	Budget available for investment to the replicating portfolio on day J
θ_{ij}	Price of instrument i in the portfolio on day j; $i = 1, \ldots, I$, $j = 1, \ldots, J$
θ_{0j}	Price of the S&P100 index on day j; $j = 1, \ldots, J$
$\mathbf{U}$	Amount of money to be on hand on final day J
$\chi = \frac{\mathbf{U}}{\theta_{0J}}$	Number of units of the S&P100 index at final day J
x_i	Number of shares of instrument i in the replicating portfolio; $i = 1, \ldots, I$
$\theta = (\theta_0, \theta_1, \ldots, \theta_I)$	Random scenario vector having J equally probable outcomes $(\theta_{0j}, \theta_{1j}, \ldots, \theta_{Ij})$; $j = 1, \ldots, J$
w	Bound on the underperformance of the replicating portfolio compared to the S&P100 index
α	Confidence level in CVaR

Main Functions

- The loss function is the relative shortfall of the tracking portfolio on day j:

$$L(x, \theta_j) = \frac{\theta_{0j}\chi - \sum_{i=1}^{I}\theta_{ij}x_i}{\theta_{0j}\chi} = 1 - \sum_{i=1}^{I}\frac{\theta_{ij}}{\theta_{0j}\chi}x_i .$$

- The *mean absolute penalty* function of the tracking portfolio is defined by

$$\textbf{meanabs\_pen}(L(x,\theta)) = \frac{1}{J}\sum_{i=1}^{J}\left|L(x,\theta_j)\right| = \frac{1}{J}\sum_{i=1}^{J}\left|1 - \sum_{i=1}^{I}\frac{\theta_{ij}}{\theta_{0j}\chi}x_i\right| .$$

- CVaR of the loss $L(x, \theta)$ is CVaR_α^- (8.6.1) implemented by the function

$$\textbf{cvar\_risk}_\alpha(L(x,\theta)) = \min_{c\in\mathbb{R}}\left(c + \frac{1}{1-\alpha}\sum_{j=1}^{J}p_j \max\left\{L_j(x,\theta) - c, 0\right\}\right),$$

  where $p_j = 1/J,\ j = 1, \ldots, J$.

Problem: problem_CS_Portfolio_Replication_CVaR_0p001

Problem 9.13. Minimize the replication error

$$\min_x \textbf{meanabs\_pen}(L(x,\theta)) \tag{9.7.1}$$

subject to the CVaR constraint on the underperformance of the replicating portfolio with respect to the S&P100 index

$$\textbf{cvar\_risk}_\alpha(L(x,\theta)) \le w, \tag{9.7.2}$$

subject to the budget constraint

$$\sum_{i=1}^{I}\theta_{iJ}x_i = U, \tag{9.7.3}$$

and subject to no-shorting constraints on exposures (numbers of shares)

$$x_i \ge 0, \qquad i = 1, \ldots, I. \tag{9.7.4}$$

Portfolio Safeguard Formulation

problem: problem_CS_Portfolio_Replication_CVaR_0p001, type = **minimize**
objective: objective_Portfolio_Replication
meanabs_pen_Portfolio_Replication(**matrix_portfolio_replication_scenarios**)
constraint: constraint_Portfolio_Replication_CVaR, **upper_bound = 0.001**
cvar_risk_Portfolio_Replication(**0.89999998, matrix_portfolio_replication_scenarios**)
constraint: constraint_Portfolio_Replication_budget, **lower_bound = 10000, upper_bound = 10000**, linearize = 0
linear_Portfolio_Replication_budget(**matrix_portfolio_replication_budget**)
box_of_variables: **lowerbounds = point_lowerbounds**
solver: VAN, precision = 3, stages = 6

Problem Datasets: Dataset1

Data	http://www.ise.ufl.edu/uryasev/files/2013/04/data_problem_CS_Portfolio_Replication_CVaR_0p001.zip
Solution	http://www.ise.ufl.edu/uryasev/files/2011/12/solution_problem_CS_Portfolio_Replication_CVaR_0p001.zip
# of variables	30
# of scenarios	600
Objective value	1.48121e–02
Running time	0.06 s (PC 2.66 GHz)

9.8 Case Study: Style Classification with Quantile Regression (pm_pen, pm_pen_g)

Background

The *return-based style classification* for a mutual fund is a regression of the fund return on several indices as explanatory variables, where regression coefficients represent the fund's style with respect to each of the indices.

Carhart (1997) and Sharpe (1992) used the *least squares linear regression* to estimate conditional expectation of the fund return distribution under the condition that a new realization of explanatory variables is observed, whereas Bassett and Chen (Bassett and Chen 2001) conducted style classification of the return distribution quantiles with the *quantile regression* suggested by Koenker and Bassett (1978).

In contrast to the least squares regression, the quantile regression can assess the impact of explanatory variables on various parts of the regressand distribution, for example, on the 95th and 99th percentiles. Moreover, for a portfolio with exposure to derivatives, the mean and quantiles of the portfolio return distribution may have quite different regression coefficients for the same explanatory variables. For example, in most cases, the strategy of investing into naked deep out-of-the-money options behaves like a bond paying some interest; however, in some rare cases, this strategy may lose significant amount of money. With the quantile regression, a fund manager can analyze the impact of a particular factor on any part of the return distribution.

This case study regresses the S&P500 index and Fidelity Magellan Fund on the Russell Value Index (RUJ), Russell 1000 Value Index (RLV), Russell 2000 Growth Index (RUO), and Russell 1000 Growth Index (RLG). The confidence level in the quantile regression varies from 0.1 to 0.9. The goal is to test whether the quantiles and the median of the residual distribution have different explanatory coefficients. Numerical experiments show that this is the case for the Fidelity Magellan Fund. In particular, the median of the Magellan distribution mostly depends on RLG and RLV funds, whereas the predictor of the 10% lower percentile of the Magellan distribution has a relatively small negative dependence on RUJ.

Quantile Regression Methodology

Let $y_1, \ldots, y_J$ be J observations (a random sample) of a random variable Y, and let $\zeta_j = (\zeta_{1j}, \ldots, \zeta_{Ij})$, $j = 1, \ldots, J$, be J corresponding observations of a covariate (explanatory) vector $\zeta = (\zeta_1, \ldots, \zeta_I)$. The estimate of the α-quantile of Y under the condition that a new realization $\hat{\zeta} = \left(\hat{\zeta}_1, \ldots, \hat{\zeta}_I\right)$ is observed is given by $\sum_{i=1}^{I} \hat{\zeta}_i \, x_i$, where the vector of regression coefficients $x = (x_1, \ldots, x_I)$ minimizes the asymmetric mean absolute error (2.0.3):

$$\min_{x \in \mathbb{R}^I} \left(\alpha \sum_{j=1}^{J} \left[y_j - \sum_{i=1}^{I} \zeta_{ij} x_i \right]_+ + (1-\alpha) \sum_{j=1}^{J} \left[\sum_{i=1}^{I} \zeta_{ij} x_i - y_j \right]_+ \right), \tag{9.8.1}$$

where $[t]_+ = \max\{t, 0\}$; see Koenker and Bassett (1978). The minimization problem (9.8.1) can be reformulated in terms of the *partial moment penalty for loss* function (**pm_pen**) and the *partial moment penalty for gain* function (**pm_pen_g**).

References

Bassett, G.W., Chen, H-L. (2001): Portfolio style: return-based attribution using quantile regression. Empirical Economics **26**, 293–305

Carhart, M.M. (1997): On persistence in mutual fund performance. Journal of Finance **52**, 57–82

Koenker, R., Bassett, G. (1978): Regression quantiles. Econometrica **46**, 33–50
Sharpe, W.F. (1992): Asset allocation: management style and performance measurement. Journal of Portfolio Management **18**(2), 7–19

Notation	Description
I	Number of style indices used for classification
$j \in \{1, \ldots, J\}$	Index of scenarios
$p_j = 1/J$	Probability of scenario j; $j = 1, \ldots, J$
θ_{0j}	Monthly rate of return of the fund, for which the classification is conducted, under scenario j
θ_{ij}	Monthly rate of return of style index i under scenario $j; i = 1, \ldots, I,\ j = 1, \ldots, J$
θ_i	Random variable assuming J equally probable values $\{\theta_{i1}, \ldots, \theta_{iJ}\}$; $i = 1, \ldots, I$
$\theta = (\theta_0, \theta_1, \ldots, \theta_I)$	Random vector
$x = (x_1, \ldots, x_I)$	Vector of regression coefficients (loading factors)
$L(x, \theta) = \theta_0 - \sum_{i=1}^{I} \theta_i x_i$	Loss function
$G(x, \theta) = \sum_{i=1}^{I} \theta_i x_i - \theta_0$	Gain function

Main Functions

The *partial moment penalty for loss* and *partial moment penalty for gain* functions are defined by

$$\textbf{pm\_pen}(L(x,\theta)) = \sum_{j=1}^{J} p_j \max\left\{0, L(x,\theta_j) - w\right\},$$

$$\textbf{pm\_pen\_g}(G(x,\theta)) = \sum_{j=1}^{J} p_j \max\left\{0, G(x,\theta_j) - w\right\},$$

respectively, where w is some target value.

Problem: problem_Style_Classification_Fidelity_Magellan_0p1

Let $\theta_{0j} = Y_j$, $\theta_{ij} = \zeta_{ij}$, and $w = 0$. Then the quantile regression (9.8.1) can be reformulated as follows.

Problem 9.14. Minimize the *asymmetric mean absolute error* for $\alpha = 0.1$:

$$\min_x \left[\alpha \; \mathbf{pm\_pen}(L(x,\theta)) + (1-\alpha) \; \mathbf{pm\_pen\_g}(G(x,\theta))\right]. \tag{9.8.2}$$

Portfolio Safeguard Formulation

Problem: problem_Style_Classification_Fidelity_Magellan_0p1, type = **minimize**
objective: objective_Magellan
0.1·pm_pen_Magellan(**0.0, matrix_style_classification_fidelity_magellan**)
+**0.9·pm_pen_g**_Magellan(**0.0, matrix_style_classification_fidelity_magellan**)
box_of_variables:
solver: VAN, precision = 5, stages = 6

Problem Datasets: Dataset1

Data	http://www.ise.ufl.edu/uryasev/files/2013/02/data_problem_pm_Style_Classification_Fidelity_Magellan_0p1.zip
Solution	http://www.ise.ufl.edu/uryasev/files/2013/02/solution_problem_pm_Style_Classification_Fidelity_Magellan_0p1.zip
# of variables	5
# of scenarios	1,264
Objective value	1.22086e–03
Running time	0.01 s (PC 2.66 GHz)

9.9 Case Study: CVaR Estimation Through Explanatory Factors with Mixed Quantile Regression (pm_pen, pm_pen_g)

Background

This case study estimates *CVaR for loss* [CVaR_α^-, see (8.6.1)] of the return distribution of a mutual fund with the *mixed quantile regression* discussed in Sect. 6.5.1. (Regression coefficients for the tail of residuals' distribution may differ from those for the mean and for the median of the distribution.) Specifically, the return of the Fidelity Magellan Fund is regressed against the returns of the Russell Value Index (RUJ), Russell 1000 Value Index (RLV), Russell 2000 Growth Index

(RUO), and Russell 1000 Growth Index (RLG). CVaR for loss with $\alpha = 0.9$ is approximated by the weighted average of four VaRs with confidence levels 0.92, 0.94, 0.96, and 0.98.

The approach is similar to the quantile regression that estimates tail quantiles of the return distribution of a mutual fund as a function of several indices (explanatory variables); see Case Study 9.8 and also Bassett and Chen (2001). Regression coefficients represent the fund's style with respect to each of the indices, and therefore, the regression is called "style classification."

Let $y_1, \ldots, y_J$ be J observations (a random sample) of a random variable Y, and let $\zeta_j = (\zeta_{1j}, \ldots, \zeta_{Ij})$, $j = 1, \ldots, J$, be J corresponding observations of a covariate (explanatory) vector $\zeta = (\zeta_1, \ldots, \zeta_I)$. The estimate of $\mathrm{CVaR}_\alpha^-(Y)$ under the condition that a new realization $\hat{\zeta} = \left(\hat{\zeta}_1, \ldots, \hat{\zeta}_I\right)$ is observed is given by $\sum_{i=1}^I \hat{\zeta}_i x_i^a$, where $x_1^a, \ldots, x_I^a$ are found from the mixed quantile regression; see Example 2.1, Example 2.6 in Rockafellar et al. (2009), and Example 10 in Rockafellar and Uryasev (2013).

References

Bassett, G.W., Chen, H-L. (2001): Portfolio style: return-based attribution using quantile regression. Empirical Economics **26**, 293–305

Rockafellar, R.T., Uryasev, S., Zabarankin, M. (2009): Risk tuning with generalized linear regression. Mathematics of Operations Research **33**(3), 712–729

Rockafellar, R.T., Uryasev, S. (2013): The fundamental risk quadrangle in risk management, optimization, and statistical estimation. Surveys in Operations Research and Management Science **18**, to appear

Notation	Description
I	Number of style indices used for classification ($I = 4$)
$i \in \{1, \ldots, I\}$	Index of style indices (i = 1, 2, 3, and 4 correspond to RLV, RLG, RUJ, and RUO, respectively)
J	Number of scenarios (time periods)
$j \in \{1, \ldots, J\}$	Index of scenarios (scenarios are equally probable)
θ_{0j}	Monthly rate of return of the Fidelity Magellan Fund (regressand) under scenario j
θ_{ij}	Monthly rate of return of style index i under scenario j; $i = 1, \ldots, I$, $j = 1, \ldots, J$
θ_i	Random variable assuming J equally probable values $\{\theta_{i1}, \ldots, \theta_{iJ}\}$; $i = 1, \ldots, I$
$\theta = (\theta_0, \theta_1, \ldots, \theta_I)$	Random vector
$x = (x_1, \ldots, x_I)$	Vector of regression coefficients (loading factors)
$L(x, \theta) = \theta_0 - \sum_{i=1}^I \theta_i x_i$	Loss function (regression residual)
$\lambda_1, \ldots, \lambda_I$	Weights in the mixed quantile regression

Main Functions

Let X be a discrete random variable assuming value X_j in scenario j with probability $p_j = 1/J$, $j = 1, \ldots, J$, and let w be some target value.

- The *partial moment penalty for loss* and *partial moment penalty for gain* functions are defined by

$$\textbf{pm\_pen}(X) = \sum_{j=1}^{J} p_j \max\left\{0, X_j - w\right\}$$

and

$$\textbf{pm\_pen\_g}(X) = \sum_{j=1}^{J} p_j \max\left\{0, w - X_j\right\},$$

respectively.

- The rescaled *asymmetric mean absolute error measure* (2.0.3) for $\alpha = \alpha_i$ is given by

$$\begin{aligned}\mathscr{E}_{\alpha_i}(X) &= \sum_{j=1}^{J} p_j \left[\frac{\alpha_i}{1-\alpha_i}\max\{0, X_j\} + \max\{0, -X_j\}\right]\\ &= \frac{\alpha_i}{1-\alpha_i}\textbf{pm\_pen}(X) + \textbf{pm\_pen\_g}(X).\end{aligned}$$

- The *mixed quantile error measure* (2.4.5) can be restated in the form

$$\begin{aligned}\mathscr{E}(X) &= \inf_{\substack{c_1,\ldots,c_I\\ \lambda_1 c_1+\cdots+\lambda_I c_I=0}} \sum_{i=1}^{I} \lambda_i \, \mathscr{E}_{\alpha_i}(X - c_i)\\ &= \min_{\substack{c_1,\ldots,c_I\\ \lambda_1 c_1+\cdots+\lambda_I c_I=0}} \sum_{i=1}^{I} \lambda_i \left[\frac{\alpha_i}{1-\alpha_i}\textbf{pm\_pen}(X - c_i) + \textbf{pm\_pen\_g}(X - c_i)\right].\end{aligned}$$

The associated statistic of $\mathscr{E}(X)$ is the mixed quantile (2.4.6); see Example 2.1 and Example 2.6 in [46].

Problem: Mixed Quantile Regression

Problem 9.15. Minimize the mixed quantile error $\mathscr{E}$ of the regression residual $L(x, \theta)$ with respect to the regression coefficient vector x and auxiliary variables $c_1, \ldots, c_I$:

$$\min_{x,\, c_1,\ldots,c_I} \sum_{i=1}^{I} \lambda_i \left[\frac{\alpha_i}{1-\alpha_i} \mathbf{pm\_pen}(L(x,\theta) - c_i) + \mathbf{pm\_pen\_g}(L(x,\theta) - c_i) \right]$$

subject to the linear constraint on $c_1, \ldots, c_I$:

$$\sum_{i=1}^{I} \lambda_i\, c_i = 0.$$

Portfolio Safeguard Formulation

Problem: problem_mixed_percentile_regression, type = **minimize**
objective: objective_Magellan
2.875·pm_pen_Magellan92(**0,"M"92)**+ **0.25·pm_pen_g**_Magellan92(**0,"M"92)**
+**3.9167·pm_pen**_Magellan94(**0,"M"94)+0.25·pm_pen_g**_Magellan94(**0,"M"94)**
+**6·pm_pen**_Magellan96(**0,"M"96)**+ **0.25·pm_pen_g**_Magellan96(**0,"M"96)**
+**12.25·pm_pen**_Magellan98(**0,"M"98)**+ **0.25·pm_pen_g**_Magellan98(**0,"M"98)**
constraint: constraint_linear, **lower_bound = 0, upper_bound = 0**
linear_matrix_of_constraint_linear(**matrix_of_constraint**)
box_of_variables:
solver: VAN, precision = 6, stages = 6

where the string "M" should be replaced by the string "matrix_style_classification_fidelity_magellan_modified."

Problem Datasets: Dataset1

Data	http://www.ise.ufl.edu/uryasev/files/2013/02/data_problem_pm_mixed_percentile_regression.zip
Solution	http://www.ise.ufl.edu/uryasev/files/2013/02/solution_problem_pm_mixed_percentile_regression.zip
# of variables	9
# of scenarios	1,264
Objective value	1.54118e–02
Running time	0.08 s (PC 2.66 GHz)

9.10 Case Study: Optimal Test Selection (cvar_comp_pos, prmulti_pen_ni_g)

Background

This case study solves three problems for optimal selection of tests subject to constraints on consumption of available resources that have limited known capacities. No partial test is allowed: each test is either conducted or not. If test resource consumptions are known and deterministic, then test selection is formulated as a *deterministic linear programming assignment problem* with boolean decision variables. If resource consumptions are uncertain, test selection is formulated in the form of *robust problem* and *stochastic problem*. The *robust problem* increases the need in each resource by 20% of its average consumption by $\alpha \cdot 100\%$ of the largest consumers, whereas the *stochastic problem* assumes that resource consumption of each test is an independent normally distributed random variable and imposes a probabilistic constraint on resource consumption. Compared to the first (deterministic) problem, the robust and stochastic problems provide more realistic solutions.

Notation	Description
I	Number of tests
$i \in \{1, \ldots, I\}$	Index of test
J	Number of resources
$j \in \{1, \ldots, J\}$	Index of resources
c_i	Value of test i
$x = (x_1, \ldots, x_I)$	Vector of boolean variables ($x_i = 1$ if test i is chosen and $x_i = 0$ otherwise)
a_{ij}	Deterministic consumption of resource j by test i in the deterministic problem
b_j	Upper bound on resource j
θ_{ij}	Random consumption of resource j by test i in the stochastic problem: all θ_{ij} are independent normally distributed, $\theta_{ij} \sim N\left(a_{ij}, 0.04a_{ij}^2\right)$
$\theta^j = (b_j, \theta_{1j}, \ldots, \theta_{Ij})$	Random vector
$G_j(x, \theta^j) = \sum_{i=1}^{I} \theta_{ij} x_i - b_j$	Gain function
d	Upper bound on the probability of at least one resource being depleted

Main Functions

- Average consumption of resource j by $\alpha \cdot 100\%$ of the largest consumers is implemented by the *CVaR component positive* function

$$\textbf{cvar\_comp\_pos}_\alpha(x,j) = \min_{c_j\in\mathbb{R}}\left(c_j + \frac{1}{n(1-\alpha)}\sum_{i=1}^{n}\max\left\{a_{ij}x_i - c_j, 0\right\}\right).$$

- If recourse consumptions are multivariate normally distributed, then the probability of at least one resource being depleted is implemented by the "*probability exceeding penalty for gain multiple normal independent*" function

$$\begin{aligned}&\textbf{prmulti\_pen\_ni\_g}\left(0, G_1(x,\theta^1),\ldots,G_J(x,\theta^J)\right)\\&= 1-\mathbb{P}\left[G_1(x,\theta^1)\le 0,\ldots,G_J(x,\theta^J)\le 0\right].\end{aligned}$$

9.10.1 Problem 1: problem_deterministic_LP_model

Problem 9.16 (Deterministic linear programming assignment model). Maximize the value of selected tests

$$\max_x \sum_{i=1}^{I} c_i x_i \tag{9.10.1}$$

subject to constraints on resources

$$\sum_{i=1}^{I} a_{ij}x_i \le b_j, \qquad j=1,\ldots,J \tag{9.10.2}$$

and subject to constraints on the decision variables

$$x_i \in \{0,1\}, \qquad i=1,\ldots,I. \tag{9.10.3}$$

Portfolio Safeguard Formulation

problem: problem_deterministic_LP_model, type = **maximize**
objective: objective_linear
linear_obj(**matrix_obj**)
constraint: constraint_MultiLin, **upper_bound = vector_right**
linearmulti_matrix_multi(**matrix_multilin**)
box_of_variables: lowerbounds = 0, upperbounds = 1, types = 1
solver: VAN, precision = 9, stages = 30

Problem Datasets: Dataset1

Data	http://www.ise.ufl.edu/uryasev/files/2013/04/data_problem_deterministic_LP_model.zip
Solution	http://www.ise.ufl.edu/uryasev/files/2011/12/solution_problem_deterministic_LP_model.zip
# of variables	21
# of scenarios	20
Objective value	879
Running time	0.05 s (PC 2.66 GHz)

9.10.2 Problem 2: problem_robust_model

Problem 9.17 (Robust model). Maximize the value of selected tests

$$\max_x \sum_{i=1}^{I} c_i x_i \tag{9.10.4}$$

subject to constraints on resources

$$\sum_{i=1}^{I} a_{ij} x_i + 0.2\ \mathbf{cvar\_comp\_pos}_{\alpha}(x, j) \le b_j, \qquad j = 1, \ldots, J, \tag{9.10.5}$$

and subject to constraints on the decision variables

$$x_i \in \{0, 1\}, \qquad i = 1, \ldots, I. \tag{9.10.6}$$

Portfolio Safeguard Formulation

problem: problem_robust_model, type = **maximize**
objective: objective_robust
linear_obj(**matrix_obj)**
constraint: constraint_c1, **upper_bound = 800**
linear_1(**matrix_c1)**
+0.2·cvar_comp_pos_1(0.80000000, matrix_c1)

⋮

constraint: constraint_ci, **upper_bound** = $a[i]$

linear_*i*(matrix_c*i*)
+0.2·cvar_comp_pos_*i*(0.80000000, matrix_c*i*)

⋮

constraint: constraint_c20, **upper_bound = 200**
linear_20(**matrix_c20)**
+0.2·cvar_comp_pos_20(0.80000000, matrix_c20)
box_of_variables: **lowerbounds = 0**, **upperbounds = 1**, types = 1
solver: CAR, precision = 9, stages = 30

where index i in the block of constraints changes from 2 to 19 and $a[i]$ is the ith element of vector $a = (800, 600, 6, 5, 6, 5, 6, 5, 6, 5, 6, 6, 9, 5, 20, 9, 20, 9, 20, 200)$.

Problem Datasets: Dataset1

Data	http://www.ise.ufl.edu/uryasev/files/2013/04/data_problem_robust_model.zip
Solution	http://www.ise.ufl.edu/uryasev/files/2011/12/solution_problem_robust_model.zip
# of variables	21
# of scenarios	20
Objective value	873
Running time	0.29 s (PC 2.66 GHz)

9.10.3 Problem 3: problem_stochastic_model

Problem 9.18 (Stochastic model). Maximize the value of selected tests

$$\max_{x} \sum_{i=1}^{I} c_i x_i \tag{9.10.7}$$

subject to a constraint on the probability of at least one resource being depleted

$$\textbf{prmulti\_pen\_ni\_g}\left(0, G_1(x, \theta^1), \ldots, G_J(x, \theta^J)\right) \leq d \tag{9.10.8}$$

and subject to constraints on the decision variables

$$x_i \in \{0, 1\}, \qquad i = 1, \ldots, I. \tag{9.10.9}$$

Portfolio Safeguard Formulation

```
problem: problem_stochastic_model, type = maximize
objective: objective_ACC
linear_obj(matrix_obj)
constraint: constraint_MultiLin, upper_bound = vector_right
linearmulti_matrix_multi(matrix_multilin)
constraint: constraint_PrMulti_NI20, upper_bound = 0.2
prmulti_pen_ni_g_20(0.0, matrix_mu_TETA, matrix_var_TETA)
box_of_variables: lowerbounds = 0, upperbounds = 1, types = 1
solver: VAN, precision = 9, stages = 30
```

Problem Datasets: Dataset1

Data	http://www.ise.ufl.edu/uryasev/files/2013/04/data_problem_stochastic_model.zip
Solution	http://www.ise.ufl.edu/uryasev/files/2011/12/solution_problem_stochastic_model.zip
# of variables	21
# of scenarios	20
Objective value	833
Running time	0.21 s (PC 2.66 GHz)

9.11 Case Study: Sparse Signal Reconstruction: A Cardinality Approach (meanabs_pen, cardn, polynom_abs)

Background

This case study formulates and solves three nonlinear optimization problems for *sparse reconstruction and compressed sensing*, in which *cardinality*[2] of a decision vector is either constrained or optimized. The first problem minimizes $\mathscr{L}_1$-error of the regression residual ($\mathscr{L}_1$-regression) subject to a constraint on cardinality of the decision vector. The second problem minimizes cardinality of the decision vector subject to a constraint on $\mathscr{L}_1$-error of the residual. The third problem solves

[2]Cardinality of a vector is the number of vector's nonzero components.

$\mathscr{L}_1$-regression subject to a constraint on the $\mathscr{L}_1$-norm[3] of the decision vector. These problems are closely related to those in Figueiredo et al. (2007), whereas details of the approach and analysis of this case study can be found in Boyko (2011).

References

Figueiredo, M., Nowak, R., Wright, S. (2007): Gradient projection for sparse reconstruction: Application to compressed sensing and other inverse problems. IEEE Journal of Selected Topics in Signal Processing **1**(4), 586–597

Boyko, N., Karamemis, G., Uryasev, S. (2011): Sparse signal reconstruction: a cardinality approach. (http://www.ise.ufl.edu/uryasev/files/2012/02/tsparse_2_1_Stan2_Gulver_2.pdf)

Notation	Description
$I + 1$	Number of columns in the data matrix
$i \in \{0, 1, \ldots, I\}$	Index of columns
J	Number of rows in the data matrix
$j \in \{1, \ldots, J\}$	Index of rows
$x = (x_1, \ldots, x_I)$	Vector of decision variables
$\theta = (\theta_0, \theta_1, \ldots, \theta_I)$	Random vector with $I + 1$ components (θ has scenarios $\theta^j = (\theta_{j0}, \theta_{j1}, \ldots, \theta_{jI})$, $j = 1, \ldots, J$, with equal probabilities $1/J$, where scenario j corresponds to row j in the data matrix)
$L(x, \theta^j) = \theta_{j0} - \sum_{i=1}^{I} \theta_{ji} x_i$	Residual corresponding to row j; $j = 1, \ldots, J$
S	Threshold on "sparsity"
ϵ	Desired accuracy for solving regression problem
u_i	Upper bound on variable x_i; $i = 1, \ldots, I$
l_i	Lower bound on variable x_i; $i = 1, \ldots, I$

Main Functions

- The *mean absolute penalty* function ($\mathscr{L}_1$-norm) is defined by

$$\textbf{meanabs\_pen}(L(x, \theta)) = \sum_{j=1}^{J} p_j \left| L(x, \theta^j) \right| ,$$

where $p_j = 1/J$, $j = 1, \ldots, J$.

[3] $\mathscr{L}_1$-norm of a vector is the sum of absolute values of vector components.

- The *cardinality* function with threshold w is defined by

$$\textbf{cardn}(x,w) = \sum_{i=1}^{I} I_{\{a_i x_i \geq w\}} + \sum_{i=1}^{I} I_{\{b_i x_i \leq -w\}},$$

where $I_{\{\cdot\}}$ is the indicator function equal to 1 if the condition in curly brackets is true and equal to 0 otherwise.

- The *polynomial absolute* function ($\mathscr{L}_1$-vector norm) is defined by

$$\textbf{polynom\_abs}(x) = \sum_{i=1}^{I} |x_i|.$$

9.11.1 Problem 1: problem_constr_cardinality

Problem 9.19. Minimize $\mathscr{L}_1$-error of the residual

$$\min_x \textbf{meanabs\_pen}(L(x,\theta)) \tag{9.11.1}$$

subject to a constraint on cardinality of the decision vector

$$\textbf{cardn}(x,w) \leq S \tag{9.11.2}$$

and subject to box constraints on the decision vector

$$l_i \leq x_i \leq u_i, \qquad i = 1,\ldots,I. \tag{9.11.3}$$

Portfolio Safeguard Formulation

problem: problem_constr_cardinality, type = **minimize**
objective: objective_meanabs_pen
meanabs_pen_matrix_objective(**matrix_objective**)
constraint: constraint_cardn, **upper_bound = 160**, linearize = 1
cardn_matrix_card(**0.0000001,matrix_card**)
box_of_variables: **lowerbounds = point_lowerbounds,**
upperbounds = point_upperbounds
solver: CAR, precision = 3, stages = 2, timelimit = 3000

Problem Datasets: Dataset1

Data	http://www.ise.ufl.edu/uryasev/files/2013/03/data_problem_constr_cardinality_no_noise.zip
Solution	http://www.ise.ufl.edu/uryasev/files/2011/12/solution_problem_constr_cardinality_no_noise.zip
# of variables	4,096
# of scenarios	1,024
Objective value	3.05777e–013
Running time	25.24 s (PC 2.66 GHz)

Problem Datasets: Dataset2

Data	http://www.ise.ufl.edu/uryasev/files/2013/03/data_problem_constr_cardinality_with_noise.zip
Solution	http://www.ise.ufl.edu/uryasev/files/2011/12/solution_problem_constr_cardinality_with_noise.zip
# of variables	4,096
# of scenarios	1,024
Objective value	3.89704e–04
Running time	21.44 s (PC 2.66 GHz)

9.11.2 Problem 2: problem_minimize_cardinality

Problem 9.20. Minimize cardinality of the decision vector

$$\min_x \mathbf{cardn}(x, w) \tag{9.11.4}$$

subject to the constraint on $\mathscr{L}_1$-error of the residual

$$\mathbf{meanabs\_pen}(L(x, \theta)) \leq \epsilon \tag{9.11.5}$$

and subject to box constraints on the decision vector

$$l_i \leq x_i \leq u_i, \qquad i = 1, \dots, I. \tag{9.11.6}$$

Portfolio Safeguard Formulation

problem: problem_minimize_cardinality, type = **minimize**
objective: objective_cardn, linearize = 1
cardn_matrix_card(**0.0000001,matrix_card**)
constraint: constraint_meanabs_pen, **upper_bound = 0.0004**
meanabs_pen_matrix_objective(**matrix_objective**)
box_of_variables: **lowerbounds = point_lowerbounds**,
upperbounds = point_upperbounds
solver: CAR, precision = 3, stages = 3, timelimit = 30000

Problem Datasets: Dataset1

Data	http://www.ise.ufl.edu/uryasev/files/2013/03/data_problem_minimize_cardinality_no_noise.zip
Solution	http://www.ise.ufl.edu/uryasev/files/2011/12/solution_problem_minimize_cardinality_no_noise.zip
# of variables	4,096
# of scenarios	1,024
Objective value	160
Running time	1888.62 s (PC 2.66 GHz)

Problem Datasets: Dataset2

Data	http://www.ise.ufl.edu/uryasev/files/2013/03/data_problem_minimize_cardinality_with_noise.zip
Solution	http://www.ise.ufl.edu/uryasev/files/2011/12/solution_problem_minimize_cardinality_with_noise.zip
# of variables	4,096
# of scenarios	1,024
Objective value	160
Running time	1613.43 s (PC 2.66 GHz)

9.11.3 Problem 3: problem_constr_polynomabs

Problem 9.21. Minimize $\mathscr{L}_1$-error of the regression residual

$$\min_x \textbf{meanabs\_pen}(L(x,\theta)) \tag{9.11.7}$$

subject to the constraint on the $\mathscr{L}_1$-norm of the decision vector

$$\textbf{polynom\_abs}(x) \le S \tag{9.11.8}$$

and subject to box constraints on the decision vector

$$l_i \le x_i \le u_i, \qquad i = 1, \ldots, I. \tag{9.11.9}$$

Portfolio Safeguard Formulation

problem: problem_constr_polynomab, type = **minimize**
objective: objective_meanabs_pen
meanabs_pen_matrix_objective(**matrix_objective)**
constraint: constraint_polynom_abs, **upper_bound = 160**, linearize = 1
polynom_abs_1(matrix_polynom_abs)
box_of_variables: **lowerbounds = point_lowerbounds**,
upperbounds = point_upperbounds
solver: CAR, precision = 3, stages = 2, timelimit = 3000

Problem Datasets: Dataset1

Data	http://www.ise.ufl.edu/uryasev/files/2013/03/data_problem_constr_polynomabs_no_noise.zip
Solution	http://www.ise.ufl.edu/uryasev/files/2011/12/solution_problem_constr_polynomabs_no_noise.zip
# of variables	4,096
# of scenarios	1,024
Objective value	6.33804e–013
Running time	20.91 s (PC 2.66 GHz)

Problem Datasets: Dataset2

Data	http://www.ise.ufl.edu/uryasev/files/2013/03/data_problem_constr_polynomabs_with_noise.zip
Solution	http://www.ise.ufl.edu/uryasev/files/2011/12/solution_problem_constr_polynomabs_with_noise.zip
# of variables	4,096
# of scenarios	1,024
Objective value	1.13060e–04
Running time	18.66 s (PC 2.66 GHz)

9.12 Case Study: Sparse Reconstruction Problems from SPARCO Toolbox (meanabs_pen, polynom_abs, linear)

Background

This case study solves several sparse reconstruction problems arising in various engineering applications including imaging, compressed sensing, geophysics, and information compressing. The objective of *sparse reconstruction* is to find a decision vector that has few nonzero components and satisfies certain linear constraints.

The problems are taken from SPARCO toolbox that offers a wide range of test problems for benchmarking of algorithms for sparse signal reconstruction; see Berg et al. (2007) and Berg and Friedlander (2008). SPARCO is implemented in MATLAB and also provides a set of linear operators for creating new test problems; see http://www.cs.ubc.ca/labs/scl/sparco/.

Typically, SPARCO toolbox problems are formulated in one of three closely related forms: *L1Relaxed*, *L1Relaxed D*, and *L2 D* (or *LASSO*). Both "L1Relaxed" and "L1Relaxed D" formulations minimize the $\mathscr{L}_1$-error of the regression residual subject to box constraints on decision variables and subject to a constraint on the $\mathscr{L}_1$-norm of the decision vector.[4] The difference in these two formulations is that "L1Relaxed D" splits each decision variable x_i into two nonnegative variables $x_i^+ = \max\{x_i, 0\}$ and $x_i^- = \max\{-x_i, 0\}$ ($x_i = x_i^+ - x_i^-$ and $|x_i| = x_i^+ + x_i^-$) and, as a result, has all decision variables nonnegative. Since "L1Relaxed D" doubles the number of the decision variables, in some problems, it may be less efficient than "L1Relaxed." The "L2 D" formulation minimizes the weighted sum of the squared $\mathscr{L}_2$-norm of the regression residual and the $\mathscr{L}_1$-norm of the vector of regression coefficients subject to box constraints on the coefficients. As the "L1Relaxed D" formulation, this one also splits each regression coefficient into two nonnegative parts.

[4] $\mathscr{L}_1$-norm of a vector is the sum of absolute values of vector components.

Small problems are solved with PSG Run-File tool, whereas large ones are solved in PSG MATLAB Environment with the PSG External Function subroutine that avoids generating full matrix and, as a result, saves time and memory.

References

Berg, E.V., Friedlander, M.P. (2008): SPARCO: A toolbox for testing sparse reconstruction algorithms. (http://www.cs.ubc.ca/labs/scl/sparco/)

Berg, E.V., Friedlander, M.P., Hennenfent, G., Herrmann, F., Saab, R., Yilmaz, O. (2007): SPARCO: A testing framework for sparse reconstruction. Tech. Rep. TR-2007-20, Dept. Computer Science, University of British Columbia, Vancouver (http://www.cs.ubc.ca/labs/scl/sparco/downloads.php?filename=sparco.pdf)

Notation	Description
I	Number of columns in the matrix of linear system
$i \in \{1, \ldots, I\}$	Index of columns
J	Number of rows in the matrix of linear system
$j \in \{1, \ldots, J\}$	Index of rows
$x = (x_1, \ldots, x_I)$	Vector of decision variables
$x_i^+ = \max\{x_i, 0\}$	Positive part of x_i
$x_i^- = \max\{-x_i, 0\}$	Positive part of $-x_i$
$x^+ = \left(x_1^+, \ldots, x_I^+\right)$	Vector of positive parts of x
$x^- = \left(x_1^-, \ldots, x_I^-\right)$	Vector of positive parts of $-x$
$\theta = (\theta_0, \theta_1, \ldots, \theta_I)$	Random vector with $I + 1$ components (θ has scenarios $\theta^j = (\theta_{j0}, \theta_{j1}, \ldots, \theta_{jI})$, $j = 1, \ldots, J$,with equal probabilities $1/J$, where scenario j corresponds to row j in the data matrix)
$L\left(x, \theta^j\right) = \theta_{j0} - \sum_{i=1}^{I} \theta_{ji} x_i$	Residual corresponding to row j; $j = 1, \ldots, J$
S	Upper bound on the sum of components of the decision vector
λ	Weight in the sum of components in the objective function in the "L2 D" formulation
$u_i \geq 0$	Upper bound on variable x_i; $i = 1, \ldots, I$
$l_i \leq 0$	Lower bound on variable x_i; $i = 1, \ldots, I$

Main Functions

- The *mean absolute penalty* function ($\mathscr{L}_1$-norm) is defined by

$$\textbf{meanabs\_pen}(L(x,\theta)) = \sum_{j=1}^{J} p_j \left|L(x,\theta^j)\right|,$$

where $p_j = 1/J$, $j = 1,\ldots,J$.
- The *polynomial absolute* function ($\mathscr{L}_1$-vector norm) is defined by

$$\textbf{polynom\_abs}(x) = \sum_{i=1}^{I} |x_i|.$$

- The *mean square penalty* function (squared $\mathscr{L}_2$-norm) is defined by

$$\textbf{meansquare}(L(x,\theta)) = \sum_{j=1}^{J} p_j \left(L(x,\theta^j)\right)^2,$$

where $p_j = 1/J$, $j = 1,\ldots,J$.
- Representation of x_i and $|x_i|$:

$$x_i = x_i^+ - x_i^-, \qquad |x_i| = x_i^+ + x_i^-, \qquad i = 1,\ldots,I,$$

where $x_i^+ \geq 0$ and $x_i^- \geq 0$, $i = 1,\ldots,I$, and, consequently,

$$L\left(x,\theta^j\right) = L\left(x^+ - x^-,\theta^j\right), \qquad \sum_{i=1}^{I} |x_i| \equiv \sum_{i=1}^{I} \left(x_i^+ + x_i^-\right).$$

9.12.1 Problem 1: L1 Relaxed

Problem 9.22 (L1 Relaxed). Minimize the $\mathscr{L}_1$-error of the regression residual

$$\min_x \textbf{meanabs\_pen}(L(x,\theta)) \tag{9.12.1}$$

subject to a constraint on the $\mathscr{L}_1$-norm of the decision vector

$$\textbf{polynom\_abs}(x) \leq S \tag{9.12.2}$$

and subject to box constraints on the decision vector

$$l_i \le x_i \le u_i, \qquad i = 1, \ldots, I. \tag{9.12.3}$$

Portfolio Safeguard Formulation (problem_601_Relaxed)

problem: problem_601_Relaxed_100, type = **minimize**
objective: objective_new, linearize = 0
meanabs_pen_obj(**matrix_ab601**)
constraint: constraint_card, **upper_bound = 100**, linearize = 1
polynom_abs_S(**matrix_card4096**)
box_of_variables: **lowerbounds = −40**, **upperbounds = 40**
solver: CAR, precision = 4, stages = 6, timelimit = 3600

Problem Datasets: Dataset1

Data	http://www.ise.ufl.edu/uryasev/files/2013/04/data_problem_601_Relaxed.zip
Solution	http://www.ise.ufl.edu/uryasev/files/2012/03/solution_problem_601_Relaxed_100.zip
# of variables	4,096
# of scenarios	3,200
Objective value	6.48
Running time	81.4 s (PC 2.83 GHz)

9.12.2 Problem 2: L1 Relaxed D

Problem 9.23 (L1 Relaxed D). Minimize the $\mathscr{L}_1$-error of the regression residual

$$\min_{x^+,\, x^-} \ \textbf{meanabs\_pen}\left(L\left(x^+ - x^-, \theta\right)\right) \tag{9.12.4}$$

subject to a constraint on the sum of components of the decision vector

$$\sum_{i=1}^{I} \left(x_i^+ + x_i^-\right) \le S \tag{9.12.5}$$

and subject to box constraints on the decision vector

$$0 \le x_i^+ \le u_i, \quad 0 \le x_i^- \le -l_i, \qquad i = 1, \ldots, I. \tag{9.12.6}$$

Portfolio Safeguard Formulation (problem_2_Relaxed_Double)

problem: problem_2_Relaxed_Double_100, type = **minimize**
objective: objective_new, linearize = 0
meanabs_pen_obj(**matrix_ab2_dbl)**
constraint: constraint_card, **upper_bound = 100**, linearize = 1
linear_S(**matrix_card1024_dbl)**
box_of_variables: **lowerbounds = 0**, **upperbounds = 1000**
solver: CAR, precision = 4, stages = 6, timelimit = 3600

Problem Datasets: Dataset1

Data	http://www.ise.ufl.edu/uryasev/files/2013/04/data_problem_2.zip
Solution	http://www.ise.ufl.edu/uryasev/files/2012/03/solution_problem_2_Relaxed_Double_100.zip
# of variables	1,024
# of scenarios	1,024
Objective value	1.22
Running time	0.5 s (PC 2.83 GHz)

9.12.3 Problem 3: L2 D

Problem 9.24 (L2 D). Minimize the weighted sum of the squared $\mathscr{L}_2$-error of the regression residual and of the $\mathscr{L}_1$-norm of the vector of regression coefficients (linear regularization part)

$$\min_{x^+,\, x^-} \left(\textbf{meansquare}\left(L\left(x^+ - x^-, \theta\right)\right) + \lambda \sum_{i=1}^{I} \left(x_i^+ + x_i^-\right) \right) \tag{9.12.7}$$

subject to box constraints on the decision vector

$$0 \le x_i^+ \le u_i, \quad 0 \le x_i^- \le -l_i, \qquad i = 1, \ldots, I. \tag{9.12.8}$$

Portfolio Safeguard Formulation (problem_2_L2_dbl)

problem: problem_2_L2_dbl_20, type = **minimize**
objective: objective_new, linearize = 0
meansquare_0(matrix_ab2_dbl)
0.039* linear_3(matrix_card1024_dbl)
constraint: constraint_card, **upper_bound = 240**, linearize = 1
polynom_abs_S(matrix_card4096)
box_of_variables: **lowerbounds = 0, upperbounds = 1000**
solver: VAN, precision = 4, stages = 2, timelimit = 3600

Problem Datasets: Dataset1

Data	http://www.ise.ufl.edu/uryasev/files/2013/04/data_problem_2.zip
Solution	http://www.ise.ufl.edu/uryasev/files/2012/03/solution_problem_2_L2_dbl_20.zip
# of variables	1,024
# of scenarios	1,024
Objective value	2.98e+03
Running time	0.32 s (PC 2.83 GHz)

9.13 Case Study: Optimal Intensity-Modulated Radiation Therapy Treatment-Planning Problem (pm2_pen)

Background

This case study solves an intensity-modulated radiation therapy (IMRT) treatment-planning problem formulated in Men et al. (2007). To penalize underdosing and overdosing with respect to a given threshold, the problem uses *quadratic one-sided penalties* or, equivalently, *second-order lower and upper partial moments*, which can be both implemented by the PSG function **pm2_pen_g**. In this case study, the involved scenario matrices are sparse (have few nonzero elements) and, therefore, can be conveniently represented in *packed matrix format* (**pmatrix**).

Reference

Men, C., Romeijn, E., Taskin, C., Dempsey, J. (2007): An exact approach to direct aperture optimization in IMRT treatment planning. Physics in Medicine and Biology **52**(24), 7333–7352

Notation	Description
$I+1$	Number of columns in the data matrix
$i \in \{0, 1, \ldots, I\}$	Index of columns
J	Number of rows in the data matrix
$j \in \{1, \ldots, J\}$	Index of rows
$x = (x_1, \ldots, x_I)$	Vector of decision variables
$\theta = (\theta_0, \theta_1, \ldots, \theta_I)$	Random vector with $I + 1$ components (θ has scenarios $\theta^j = (\theta_{j0}, \theta_{j1}, \ldots, \theta_{jI})$, $j = 1, \ldots, J$, with equal probabilities $1/J$, where scenario j corresponds to row j in the data matrix)
$L(x, \theta^j) = \theta_{j0} - \sum_{i=1}^{I} \theta_{ji} x_i$	Residual corresponding to row j; $j = 1, \ldots, J$
$\omega = 0$	Threshold value for the second-order partial moment penalty function
u_i	Upper bound on variable x_i; $i = 1, \ldots, I$
l_i	Lower bound on variable x_i; $i = 1, \ldots, I$

Main Functions

The *second-order partial moment penalty* function is defined by

$$\mathbf{pm2\_pen\_g}(L(x,\theta)) = \sum_{j=1}^{J} p_j \left(\max\left\{0, -L(x,\theta^j) - \omega\right\}\right)^2,$$

where $p_j = 1/J$, $j = 1, \ldots, J$.

Problem: problem_men

Problem 9.25. Minimize the second-order partial moment penalty function

$$\min_x \ \mathbf{pm2\_pen\_g}(L(x,\theta)) \tag{9.13.1}$$

subject to box constraints on the decision variables

$$l_i \leq x_i \leq u_i, \qquad i = 1, \ldots, I. \tag{9.13.2}$$

The following four PSG formulations of the problem (9.13.1)–(9.13.2) differ only in the datasets that they use, namely, Dataset1, Dataset2, Dataset3, and Dataset4. Dataset1 has a purely demonstrative purpose, whereas Dataset2, Dataset3, and Dataset4 are of large size and from real-life problems.

Portfolio Safeguard Formulation (problem_men_short)

problem: problem_men_short, type = **minimize**
objective: objective_men1
pm2_pen_g_men_short(**0.0, matrix_men_shell**)
box_of_variables: **lowerbounds = point_lowerbounds**,
upperbounds = point_upperbounds
solver: BULDOZER, precision = 9, stages = 6

Problem Datasets: Dataset1

Data	http://www.ise.ufl.edu/uryasev/files/2013/04/data_problem_men_short.zip
Solution	http://www.ise.ufl.edu/uryasev/files/2011/12/solution_problem_men_short.zip
# of variables	10
# of scenarios	10
Objective value	3.50098e–04
Running time	0.02 s (PC 2.83 GHz)

Portfolio Safeguard Formulation (problem_men1)

problem: problem_men1, type = **minimize**
objective: objective_men1
pm2_pen_g_men(**0.0, pmatrix_ftest_men1**)
box_of_variables: **lowerbounds = point_lowerbounds_1**,
upperbounds = point_upperbounds_1
solver: BULDOZER, precision = 9, stages = 6, timelimit = 25,
init_point = point_initial_men1

Problem Datasets: Dataset2

Data	http://www.ise.ufl.edu/uryasev/files/2013/04/data_problem_men1.zip
Solution	http://www.ise.ufl.edu/uryasev/files/2011/12/solution_problem_men1.zip
# of variables	1,113
# of scenarios	20,859
Objective value	15.6578
Running time	25.16 s (PC 2.83 GHz)

Portfolio Safeguard Formulation (problem_men2)

problem: problem_men2, type =**minimize**
objective: objective_men2
pm2_pen_g_men(**0.0, pmatrix_ftest_men2)**
box_of_variables: **lowerbounds = point_lowerbounds_2**,
upperbounds = point_upperbounds_2
solver: BULDOZER, precision = 9, timelimit = 50,
init_point = point_initial_men2

Problem Datasets: Dataset3

Data	http://www.ise.ufl.edu/uryasev/files/2013/04/data_problem_men2.zip
Solution	http://www.ise.ufl.edu/uryasev/files/2011/12/solution_problem_men2.zip
# of variables	2,736
# of scenarios	44,362
Objective value	5.61869
Running time	50.3 s (PC 2.83 GHz)

Portfolio Safeguard Formulation (problem_men3)

problem: problem_men3, type = **minimize**
objective: objective_men3
pm2_pen_g_men(**0.0, pmatrix_ftest_men3)**
box_of_variables: **lowerbounds = point_lowerbounds_3,**
upperbounds = point_upperbounds_3
solver: BULDOZER, precision = 6, stages = 6, timelimit = 50,
init_point = point_initial_men3

Problem Datasets: Dataset4

Data	http://www.ise.ufl.edu/uryasev/files/2013/04/data_problem_men3.zip
Solution	http://www.ise.ufl.edu/uryasev/files/2011/12/solution_problem_men3.zip
# of variables	2,055
# of scenarios	37,794
Objective value	3.93620
Running time	50.21 s (PC 2.83 GHz)

9.14 Case Study: Portfolio Optimization—CVaR Versus Standard Deviation (cvar_dev, st_dev)

Background

This case study solves three single-period *portfolio selection* problems that minimize portfolio risk subject to constraints on portfolio budget and portfolio expected rate of return, and subject to no-shorting constraints. The problems differ only in how the risk is measured. Let a random variable X be the portfolio rate of return. The first and second problems minimize standard deviation of X: the first one uses given covariance matrix of instruments' rates of return, whereas the second uses a matrix of scenarios for instruments' rates of return. Both are the Markowitz problem (8.2.9). The third problem minimizes CVaR deviation of X modeled with the matrix of scenarios from the second problem. It is formulated by (8.2.14) with $\mathscr{D} = \mathrm{CVaR}_\alpha^\Delta$.

Notation	Description
I	Number of instruments (clusters) in the portfolio
$i \in \{1, \ldots, I\}$	Index of instruments in the portfolio
J	Number of scenarios
$j \in \{1, \ldots, J\}$	Index of scenarios
$x = (x_1, \ldots, x_I)$	Decision vector (weights of instruments in the portfolio)
r_i	Rate of return (random variable) of instrument i
$r = (r_1, \ldots, r_I)$	Random vector
r_{ij}	Rate of return of instrument i under scenario j
Λ	Covariance matrix built based on the matrix of scenarios
α	Confidence level in CVaR deviation
$P(x,r) = \sum_{i=1}^{I} r_i x_i$	Portfolio rate of return
$P_j(x) = \sum_{i=1}^{I} r_{ij} x_i$	Portfolio rate of return under scenario j
μ	Lower bound on the portfolio expected rate of return
$L(x,r) = -P(x,r)$	Portfolio loss function
$L_j(x) = -P_j(x)$	Portfolio loss function under scenario j

Main Functions

- Standard deviation of the portfolio loss $L(x,r)$ is defined by

$$\textbf{st\_dev}(L(x,r)) = \sqrt{\sum_{j=1}^{J} p_j \left(L_j(x) - E[L(x,r)]\right)^2},$$

where

$$E[L(x,r)] = \sum_{j=1}^{J} p_j L_j(x).$$

When the covariance matrix Λ of rates of return $r_1, \ldots, r_I$ is given, **st_dev** is defined by

$$\textbf{st\_dev}\,(x, \Lambda) = \sqrt{\sum_{i=1}^{I} \sum_{k=1}^{I} \Lambda_{ik}\, x_i x_k}.$$

- *CVaR deviation* of the portfolio loss $L(x,r)$ is implemented by the function

$$\textbf{cvar\_dev}_\alpha(L(x,r)) = \textbf{cvar\_risk}_\alpha\,(L(x,r) - E[L(x,r)])\,,$$

where **cvar_risk** is *CVaR for loss* [CVaR_α^-, see (8.6.1)]:

$$\mathbf{cvar\_risk}_\alpha(L(x,r)) = \min_{c\in\mathbb{R}} \left(c + \frac{1}{1-\alpha} \sum_{j=1}^{J} p_j \max\left\{L_j(x) - c, 0\right\} \right).$$

9.14.1 Problem 1: problem_st_dev_covariances_2p9

Problem 9.26 (Markowitz problem: standard deviation with covariance matrix). Minimize standard deviation of the portfolio loss function

$$\min_x \mathbf{st\_dev}\,(x, \Lambda) \tag{9.14.1}$$

subject to the budget constraint

$$\sum_{i=1}^{I} x_i = 1, \tag{9.14.2}$$

a constraint on the portfolio expected rate of return

$$\sum_{i=1}^{I} E[r_i]\, x_i \geq \mu, \tag{9.14.3}$$

and subject to no-shorting constraints

$$x_i \geq 0, \qquad i = 1, \ldots, I. \tag{9.14.4}$$

Portfolio Safeguard Formulation

problem: problem_st_dev_covariances_2p9, type = **minimize**
objective: objective_st_dev_covariances
st_dev_covariances(**smatrix_covariances)**
constraint: constraint_budget, **lower_bound = 1**, **upper_bound = 1**, linearize = 0
linear_budget(**matrix_budget)**
constraint: constraint_return, **lower_bound = 2.9e–2**, linearize = 0
linear_returns(**matrix_returns)**
box_of_variables: **lowerbounds = point_lowerbounds**
solver: VAN, precision = 6

Problem Datasets: Dataset1

Data	http://www.ise.ufl.edu/uryasev/files/2013/04/data_problem_st_dev_covariances_2p9.zip
Solution	http://www.ise.ufl.edu/uryasev/files/2011/12/solution_problem_st_dev_covariances_2p9.zip
# of variables	10
# of scenarios	100
Objective value	8.74964e–03
Running time	0.01 s (PC 2.66 GHz)

9.14.2 Problem 2: problem_st_dev_scenarios_2p9

Problem 9.27 (Markowitz problem: standard deviation with the matrix of scenarios). Minimize standard deviation of the portfolio loss function

$$\min_{x} \textbf{st\_dev}\,(L(x,r)) \tag{9.14.5}$$

subject to the budget constraint

$$\sum_{i=1}^{I} x_i = 1, \tag{9.14.6}$$

a constraint on the portfolio expected rate of return

$$\sum_{i=1}^{I} E[r_i]\, x_i \geq \mu, \tag{9.14.7}$$

and subject to no-shorting constraints

$$x_i \geq 0, \qquad i = 1, \ldots, I. \tag{9.14.8}$$

Portfolio Safeguard Formulation

problem: problem_st_dev_scenarios_2p9, type = **minimize**
objective: objective_st_dev_scenarios
st_dev_scenarios(**matrix_scenarios**)
constraint: constraint_budget, **lower_bound = 1**, **upper_bound = 1**, linearize = 0
linear_budget(**matrix_budget**)
constraint: constraint_return, **lower_bound = 2.9e−2**, linearize = 0
linear_returns(**matrix_returns**)
box_of_variables: **lowerbounds = point_lowerbounds**
solver: BULDOZER, precision = 6

Problem Datasets: Dataset1

Data	http://www.ise.ufl.edu/uryasev/files/2013/04/data_problem_st_dev_scenarios_2p9.zip
Solution	http://www.ise.ufl.edu/uryasev/files/2011/12/solution_problem_st_dev_scenarios_2p9.zip
# of variables	10
# of scenarios	1,000
Objective value	8.74963e–03
Running time	0.01 s (PC 2.66 GHz)

9.14.3 *Problem 3: problem_min_cvar_dev_2p9*

Problem 9.28 (Markowitz-type portfolio selection problem with CVaR deviation). Minimize CVaR deviation of the portfolio loss function with $\alpha = 0.99$

$$\min_x \mathbf{cvar\_dev}_\alpha(L(x,r)) \tag{9.14.9}$$

subject to the budget constraint

$$\sum_{i=1}^{I} x_i = 1, \tag{9.14.10}$$

a constraint on the portfolio expected rate of return

$$\sum_{i=1}^{I} E[r_i]\, x_i \geq \mu, \tag{9.14.11}$$

and subject to no-shorting constraints

$$x_i \geq 0, \qquad i = 1, \ldots, I. \tag{9.14.12}$$

Portfolio Safeguard Formulation

problem: problem_min_cvar_dev_2p9, type = **minimize**
objective: objective_cvar_dev
cvar_dev_loss(**0.99, matrix_scenarios)**
constraint: constraint_budget, **lower_bound = 1**, **upper_bound = 1**, linearize = 0
linear_budget(**matrix_budget)**
constraint: constraint_return, **lower_bound = 2.9e−2**, linearize = 0
linear_returns(**matrix_returns)**
box_of_variables: lowerbounds = point_lowerbounds
solver: VAN, precision = 6

Problem Datasets: Dataset1

Data	http://www.ise.ufl.edu/uryasev/files/2013/04/data_problem_min_cvar_dev_2p9.zip
Solution	http://www.ise.ufl.edu/uryasev/files/2011/12/solution_problem_min_cvar_dev_2p9.zip
# of variables	10
# of scenarios	1,000
Objective value	3.63177e–02
Running time	0.01 s (PC 2.66 GHz)

9.15 Case Study: Portfolio Credit-Risk Optimization Modeled by Scenarios and Mixtures of Normal Distributions

Background

This case study solves six credit-risk minimization problems for a portfolio of financial contracts with different counterparties. The problems exploit conditional independence of counterparties under a structural credit-risk model and approximate the portfolio loss distribution by either "conventional" scenarios or a mixture of normal distributions. In all problems, the portfolio is optimized subject to the same constraints on budget and on portfolio expected rate of return and with the same bounds on portfolio weights. The problems differ in two ways: how the portfolio loss distribution is approximated and how the risk is measured. Namely, Problems 1 and 2 use "conventional" scenarios, whereas Problems 3–6 use mixtures of independent normal distributions. Then Problems 1, 3, and 5 minimize VaR of the portfolio loss, whereas Problems 2, 4, and 6 minimize CVaR. Problems 5 and 6 are, in fact, reformulations of Problems 3 and 4, respectively, with VaR and CVaR being represented through the CVaR optimization formula (1.4.4); see Rockafellar and Uryasev (2000). Details of the credit-risk setting can be found in Iscoe et al. (2012).

References

Iscoe, I., Kreinin, A., Mausser, H., Romanko, A. (2012): Portfolio credit-risk optimization. Journal of Banking and Finance **36**(6), 1604–1615

Rockafellar, R.T., Uryasev, S. (2000): Optimization of conditional value-at-risk. Journal of Risk **2**, 21–41

Notation	Description
J	Number of scenarios
$j \in \{1, \ldots, J\}$	Index of scenarios
$p_j = 1/J$	Probability of scenario j, $j = 1, \ldots, J$
I	Number of stocks
$i \in \{1, \ldots, I\}$	Index of stocks in the portfolio
x_i	Fraction (weight) of capital invested in stock i
$x = (x_1, \ldots, x_I)$	Vector of portfolio weights
r_{ij}	Random rate of return of stock i in scenario j (r_{ij} has mean μ_{ij} and variance v_{ij})
$P_j(x) = \sum_{i=1}^{I} r_{ij}\, x_i$	Portfolio random rate of return in scenario j
$L_j(x) = -E[P_j(x)]$	Loss function in scenario j in Problems 1–4
$\sigma_j(x) = \sigma\left(P_j(x)\right)$	Standard deviation of $P_j(x)$
$L(x)$	Random loss function with discrete probability distribution $\mathbb{P}[L(x) = L_j(x)] = p_j$, $j = 1, \ldots, J$
x_{var}	Auxiliary variable equal to VaR of $L(x)$ for optimal x in Problems 5 and 6
$x' = (x_{\text{var}}, x_1, \ldots, x_I)$	Vector of decision variables in Problems 5 and 6
$L'_j(x') = x_{\text{var}} - L_j(x)$	Loss function in scenario j in Problems 5 and 6
$L'(x')$	Random loss function with discrete probability distribution $\mathbb{P}[L'(x') = L'_j(x')] = p_j$, $j = 1, \ldots, J$
$x^0 = \left(x_1^0, \ldots, x_I^0\right)$	Positions in the initial portfolio
α	Confidence level in VaR and CVaR

Note 9.1. $L(x)$ should not be confused with $(L_1(x), \dots, L_J(x))$. The former is defined as a random variable with the discrete probability distribution $\mathbb{P}[L(x) = L_j(x)] = p_j$, $j = 1, \dots, J$, whereas the latter is a deterministic vector.

Main Functions

- The functions $L_j(x)$ and $\sigma_j^2(x)$ are determined by

$$L_j(x) = -\sum_{i=1}^{I} \mu_{ij} x_i, \qquad \sigma_j^2(x) = \sum_{i=1}^{I} v_{ij} x_i^2, \qquad j = 1, \dots, J.$$

- **var_risk**$_\alpha(L(x))$ is *VaR for loss* defined by

$$\textbf{var\_risk}_\alpha(L(x)) = \arg\min_{c \in \mathbb{R}} \left(c + \frac{1}{1-\alpha} \sum_{j=1}^{J} p_j \max\left\{L_j(x) - c, 0\right\} \right).$$

- **cvar_risk**$_\alpha(L(x))$ is *CVaR for loss* [CVaR$_\alpha^-$, see (8.6.1)] defined by

$$\textbf{cvar\_risk}_\alpha(L(x)) = \min_{c \in \mathbb{R}} \left(c + \frac{1}{1-\alpha} \sum_{j=1}^{J} p_j \max\left\{L_j(x) - c, 0\right\} \right).$$

- Let $\phi(t) = \exp\left(-t^2/2\right) / \sqrt{2\pi}$ and $\Phi(t) = \int_{-\infty}^{t} \phi(s)\, ds$ be the PDF and CDF of the standard normal distribution $N(0,1)$, respectively. The PDF of the loss L is the mixture of PDFs $\frac{1}{\sigma_j(x)} \phi\left(\frac{t - L_j(x)}{\sigma_j(x)}\right)$, $j = 1, \dots, J$, with scenario probabilities $p_j > 0$, $j = 1, \dots, J$:

$$f_L(t) = \sum_{j=1}^{J} \frac{p_j}{\sigma_j(x)} \phi\left(\frac{t - L_j(x)}{\sigma_j(x)}\right),$$

  so that the CFD of L is given by

$$F_L(\tau) = \int_{-\infty}^{\tau} f_L(t)\, dt = \sum_{j=1}^{J} p_j\, \Phi\left(\frac{\tau - L_j(x)}{\sigma_j(x)}\right).$$

  Let $F_L^{-1}(\alpha)$ be the inverse of $F_L(\tau)$ for $\alpha \in (0,1)$, and let $\Theta(\tau)$ be the first-order upper partial moment of the standard normal distribution $N(0,1)$:

$$\Theta(\tau) = \int_{\tau}^{\infty} (t-\tau)\phi(t)\,dt = \phi(\tau) - \tau\,(1-\Phi(\tau))\,.$$

- **avg_pr_pen_ni**$_\tau$ $(L_1(x),\ldots,L_J(x))$ is the *average probability exceeding penalty for loss normal independent* function with level τ, which is $1 - F_L(\tau)$, and is defined by

$$\textbf{avg\_pr\_pen\_ni}_\tau\,(L_1(x),\ldots,L_J(x)) = \sum_{j=1}^{J} p_j\,\textbf{pr\_pen\_ni}_\tau\,\big(L_j(x)\big)\,,$$

where

$$\textbf{pr\_pen\_ni}_\tau\,\big(L_j(x)\big) = 1 - \Phi\left(\frac{\tau - L_j(x)}{\sigma_j(x)}\right).$$

- **avg_pm_pen_ni**$_\tau$ $(L_1(x),\ldots,L_J(x))$ is the *average partial moment penalty for loss normal independent* function with level τ, which is $\int_\tau^\infty (t-\tau)\,f_L(t)\,dt$, and is defined by

$$\textbf{avg\_pm\_pen\_ni}_\tau\,(L_1(x),\ldots,L_J(x)) = \sum_{j=1}^{J} p_j\,\textbf{pm\_pen\_ni}_\tau\,\big(L_j(x)\big)\,,$$

where

$$\textbf{pm\_pen\_ni}_\tau\,\big(L_j(x)\big) = \sigma_j(x)\,\Theta\left(\frac{\tau - L_j(x)}{\sigma_j(x)}\right).$$

- **avg_var_risk_ni**$_\alpha$ $(L_1(x),\ldots,L_J(x))$ is the *average value-at-risk for loss normal independent* function with confidence level α, which is $F_L^{-1}(1-\alpha)$, and is defined by

$$\begin{aligned}&\textbf{avg\_var\_risk\_ni}_\alpha\,(L_1(x),\ldots,L_J(x))\\ &= \min_\tau\,\{\tau\,|\textbf{avg\_pr\_pen\_ni}_\tau\,(L_1(x),\ldots,L_J(x)) \le 1-\alpha\}\,.\end{aligned}$$

- **avg_cvar_risk_ni**$_\alpha$ $(L_1(x),\ldots,L_J(x))$ is the *average CVaR for loss normal independent* function with confidence level α, which is $\frac{1}{1-\alpha}\int_\alpha^1 F_L^{-1}(s)\,ds \equiv \frac{1}{1-\alpha}\int_{F_L^{-1}(\alpha)}^\infty t\,f_L(t)\,dt \equiv \min_{\tau\in\mathbb{R}}\big(\tau + \frac{1}{1-\alpha}\int_\tau^\infty (t-\tau)\,f_L(t)\,dt\big)$ and is defined by

$$\begin{aligned}&\textbf{avg\_cvar\_risk\_ni}_\alpha\,(L_1(x),\ldots,L_J(x))\\ &= \min_{\tau\in\mathbb{R}}\left(\tau + \frac{1}{1-\alpha}\textbf{avg\_pm\_pen\_ni}_\tau\,(L_1(x),\ldots,L_J(x))\right).\end{aligned}$$

Equivalent reformulations:

$$\mathbf{avg\_pr\_pen\_ni}_0(L'_1(x'), \ldots, L'_J(x')) = \mathbf{avg\_pr\_pen\_ni}_{x_{\text{var}}}(L_1(x), \ldots, L_J(x)),$$

$$\mathbf{avg\_pm\_pen\_ni}_0(L'_1(x'), \ldots, L'_J(x')) = \mathbf{avg\_pm\_pen\_ni}_{x_{\text{var}}}(L_1(x), \ldots, L_J(x)).$$

9.15.1 Problem 1: problem_var_LLN

Problem 9.29 (Value-at-risk). Minimize VaR of the random loss function $L(x)$

$$\min_x \mathbf{var\_risk}_\alpha(L(x)) \tag{9.15.1}$$

subject to the budget constraint

$$\sum_{i=1}^{I} x_i = 1, \tag{9.15.2}$$

a constraint on the portfolio expected rate of return

$$\sum_{i=1}^{I} \left(\sum_{j=1}^{J} p_j \mu_{ij} \right) x_i \geq \mu, \tag{9.15.3}$$

and subject to box constraints on the decision variables

$$0 \leq x_i \leq 2x_i^0, \qquad i = 1, \ldots, I. \tag{9.15.4}$$

Portfolio Safeguard Formulation

problem: problem_var_LLN, type = **minimize**
objective: objective_cp_var
var_risk_obj(**0.999, matrix_cp_mean**)
constraint: constraint_cp_budget, **lower_bound = 1**, **upper_bound = 1**
linear_budget(**matrix_cp_budget**)
constraint: constraint_cp_return, **lower_bound = 0.0963**
linear_return(**matrix_cp_return**)
box_of_variables: **lowerbounds = 0**, **upperbounds = point_2x0**
solver: tank, precision = 5, init_point = point_x0, timelimit = 600

Problem Datasets: Dataset1

Data	http://www.ise.ufl.edu/uryasev/files/2013/04/data_problems_1_4.zip
Solution	http://www.ise.ufl.edu/uryasev/files/2011/12/solution_problem_var_LLN.zip
# of variables	3,000
# of scenarios	10,000
Objective value	2.80863
Running time	22.75 s (PC 2.66 GHz)

9.15.2 Problem 2: problem_cvar_LLN

Problem 9.30 (Conditional value-at-risk). Minimize CVaR of the random loss function $L(x)$

$$\min_x \textbf{cvar\_risk}_\alpha(L(x)) \tag{9.15.5}$$

subject to the budget constraint

$$\sum_{i=1}^{I} x_i = 1, \tag{9.15.6}$$

a constraint on the portfolio expected rate of return

$$\sum_{i=1}^{I} \left(\sum_{j=1}^{J} p_j \mu_{ij} \right) x_i \geq \mu, \tag{9.15.7}$$

and subject to box constraints on the decision variables

$$0 \leq x_i \leq 2x_i^0, \qquad i = 1, \ldots, I. \tag{9.15.8}$$

Portfolio Safeguard Formulation

problem: problem_cvar_LLN, type = **minimize**
objective: objective_cp_cvar
cvar_risk_obj(**0.999, matrix_cp_mean)**
constraint: constraint_cp_budget, **lower_bound = 1**, **upper_bound = 1**
linear_budget(**matrix_cp_budget)**

constraint: constraint_cp_return, **lower_bound = 0.0963**
linear_return(**matrix_cp_return**)
box_of_variables: **lowerbounds = 0**, **upperbounds = point_2x0**
solver: tank, precision = 5, init_point = point_x0, timelimit = 600

Problem Datasets: Dataset1

Data	http://www.ise.ufl.edu/uryasev/files/2013/04/data_problems_5_6.zip
Solution	http://www.ise.ufl.edu/uryasev/files/2011/12/solution_problem_cvar_LLN.zip
# of variables	3,000
# of scenarios	10,000
Objective value	3.41393
Running time	2.065 s (PC 2.66 GHz)

9.15.3 *Problem 3: problem_avg_var_CLT*

Problem 9.31 (Average value-at-risk). Minimize VaR of the random loss $L(x)$, whose distribution is modeled by a mixture of normal independent distributions,

$$\min_x \textbf{avg\_var\_risk\_ni}_\alpha \left(L_1(x), \ldots, L_J(x)\right) \tag{9.15.9}$$

subject to the budget constraint

$$\sum_{i=1}^{I} x_i = 1, \tag{9.15.10}$$

a constraint on the portfolio expected rate of return

$$\sum_{i=1}^{I} \left(\sum_{j=1}^{J} p_j \mu_{ij} \right) x_i \geq \mu, \tag{9.15.11}$$

and subject to box constraints on the decision variables

$$0 \leq x_i \leq 2x_i^0, \qquad i = 1, \ldots, I. \tag{9.15.12}$$

Portfolio Safeguard Formulation

problem: problem_avg_var_CLT, type = **minimize**
objective: objective_cp_avg_var
+1 * **avg_var_risk_ni**_obj(**0.999, matrix_cp_mean, matrix_cp_var)**
constraint: constraint_cp_budget, **lower_bound = 1**, **upper_bound = 1**
linear_budget(**matrix_cp_budget)**
constraint: constraint_cp_return, **lower_bound = 0.0963**
linear_return(**matrix_cp_return)**
box_of_variables: **lowerbounds = 0**, **upperbounds = point_2x0**
solver: tank, precision = 5, init_point = point_x0, timelimit = 3600

Problem Datasets: Dataset1

Data	http://www.ise.ufl.edu/uryasev/files/2013/04/data_problems_1_4.zip
Solution	http://www.ise.ufl.edu/uryasev/files/2011/12/solution_problem_avg_var_CLT.zip
# of variables	3,000
# of scenarios	10,000
Objective value	3.13763
Running time	362.7 s (PC 2.66 GHz)

9.15.4 Problem 4: problem_avg_cvar_CLT

Problem 9.32 (Average conditional value-at-risk). Minimize CVaR of the random loss $L(x)$, whose distribution is modeled by a mixture of normal independent distributions,

$$\min_x \textbf{avg\_cvar\_risk\_ni}_\alpha\, (L_1(x), \ldots, L_J(x)) \tag{9.15.13}$$

subject to the budget constraint

$$\sum_{i=1}^{I} x_i = 1, \tag{9.15.14}$$

a constraint on the portfolio expected rate of return

$$\sum_{i=1}^{I} \left(\sum_{j=1}^{J} p_j \mu_{ij} \right) x_i \geq \mu, \tag{9.15.15}$$

and subject to box constraints on the decision variables

$$0 \leq x_i \leq 2x_i^0, \qquad i = 1, \ldots, I. \tag{9.15.16}$$

Portfolio Safeguard Formulation

problem: problem_avg_cvar_CLT, type = **minimize**
objective: objective_cp_avg_cvar
+1 * **avg_cvar_risk_ni**_obj(**0.999, matrix_cp_mean, matrix_cp_var)**
constraint: constraint_cp_budget, **lower_bound = 1**, **upper_bound = 1**
linear_budget(**matrix_cp_budget)**
constraint: constraint_cp_return, **lower_bound = 0.0963**
linear_return(**matrix_cp_return)**
box_of_variables: **lowerbounds = 0**, **upperbounds = point_2x0**
solver: tank, precision = 5, init_point = point_x0, timelimit = 3600

Problem Datasets: Dataset1

Data	http://www.ise.ufl.edu/uryasev/files/2013/04/data_problems_1_4.zip
Solution	http://www.ise.ufl.edu/uryasev/files/2011/12/solution_problem_avg_cvar_CLT.zip
# of variables	3,000
# of scenarios	10,000
Objective value	3.64890
Running time	296.1 s (PC 2.66 GHz)

9.15.5 Problem 5: problem_avg_var_CLT (Alternative Formulation)

Problem 9.33 (Average value-at-risk: alternative formulation). Minimize VaR of the random loss $L(x)$, whose distribution is modeled by a mixture of normal independent distributions,

$$\min_{x'} \; x_{\text{var}} \tag{9.15.17}$$

subject to the probability constraint

$$\textbf{avg\_pr\_pen\_ni}_0 \left(L'_1(x'), \ldots, L'_J(x') \right) \leq 1 - \alpha, \tag{9.15.18}$$

the budget constraint

$$\sum_{i=1}^{I} x_i = 1, \tag{9.15.19}$$

a constraint on the portfolio expected rate of return

$$\sum_{i=1}^{I} \left(\sum_{j=1}^{J} p_j \mu_{ij} \right) x_i \geq \mu, \tag{9.15.20}$$

and subject to box constraints on the decision variables

$$0 \leq x_i \leq 2x_i^0, \qquad i = 1, \ldots, I. \tag{9.15.21}$$

Portfolio Safeguard Formulation

problem: problem_avg_var_clt_alt, type = **minimize**
objective: objective_cp_obj
+1 * **variable_xvar(xvar)**
constraint: constraint_avgpr, **upper_bound = 0.001**
+1 * **avg_pr_pen_ni_**1**(0.0, matrix_cp_mean_xvar, matrix_cp_var_xvar)**
constraint: constraint_cp_budget, **lower_bound = 1**, **upper_bound = 1**
linear_budget(**matrix_cp_budget)**
constraint: constraint_cp_return, **lower_bound = 0.0963**
linear_return(**matrix_cp_return)**
box_of_variables: **lowerbounds = 0**, **upperbounds = point_2x0**
solver: tank, precision = 5, init_point = point_x0, timelimit = 600

Problem Datasets: Dataset1

Data	http://www.ise.ufl.edu/uryasev/files/2013/04/data_problems_5_6.zip
Solution	http://www.ise.ufl.edu/uryasev/files/2011/12/solution_problem_avg_var_CLT_alt.zip
# of variables	3,001
# of scenarios	10,000
Objective value	3.13763
Running time	299.47 s (PC 2.66 GHz)

9.15.6 Problem 6: problem_avg_cvar_CLT (Alternative Formulation)

Problem 9.34 (Average conditional value-at-risk: alternative formulation). Minimize CVaR of the random loss $L(x)$, whose distribution is modeled by a mixture of normal independent distributions,

$$\min_{x'} \left(x_{\text{var}} + \frac{1}{1-\alpha} \, \mathbf{avg\_pm\_pen\_ni}_0 \left(L'_1(x'), \ldots, L'_J(x') \right) \right) \tag{9.15.22}$$

subject to the budget constraint

$$\sum_{i=1}^{I} x_i = 1, \tag{9.15.23}$$

a constraint on the portfolio expected rate of return

$$\sum_{i=1}^{I} \left(\sum_{j=1}^{J} p_j \mu_{ij} \right) x_i \geq \mu, \tag{9.15.24}$$

and subject to box constraints on the decision variables

$$0 \leq x_i \leq 2x_i^0, \qquad i = 1, \ldots, I. \tag{9.15.25}$$

Portfolio Safeguard Formulation (same as for Problem 4)

---

problem: problem_avg_cvar_CLT, type = **minimize**
objective: objective_cp_avg_cvar
+1 * **avg_cvar_risk_ni**_obj(**0.999, matrix_cp_mean, matrix_cp_var)**
constraint: constraint_cp_budget, **lower_bound = 1, upper_bound = 1**
linear_budget(**matrix_cp_budget)**
constraint: constraint_cp_return, **lower_bound = 0.0963**
linear_return(**matrix_cp_return)**
box_of_variables: lowerbounds = 0, upperbounds = point_2x0
solver: tank, precision = 5, init_point = point_x0, timelimit = 3600

Problem Datasets: Dataset1

Data	http://www.ise.ufl.edu/uryasev/files/2013/04/data_problems_5_6.zip
Solution	http://www.ise.ufl.edu/uryasev/files/2012/02/solution_problem_avg_cvar_CLT_alt.zip
# of variables	3,001
# of scenarios	10,000
Objective value	3.64890
Running time	237.09 s (PC 2.66 GHz)

9.16 Case Study: ν-SVM with VaR and CVaR

Background

This case study demonstrates the *CVaR* methodology in application to *support vector machine* (SVM) classification.

Given training data $(\xi_1, y_1), \ldots, (\xi_J, y_J)$ from two categories of observations, where $\xi_j \in \mathbb{R}^n$ is a vector of features and $y_j \in \{-1, 1\}$ is a category label ($y_j = 1$ if observation j belongs to one category, and $y_j = -1$ if it belongs to the other), an *SVM aims to separate the features from the two categories by a hyperplane with maximal margin*. The hyperplane is determined by $w^\top \xi + b = 0$, where $w \in \mathbb{R}^n$ and $b \in \mathbb{R}$ are parameters, and the features are separated if $w^\top \xi_j + b \geq 0$ for $y_j = 1$ and $w^\top \xi_j + b \leq 0$ for $y_j = -1$ or simply if $y_j(w^\top \xi_j + b) \geq 0$ for all j. There are several SVM problems:

- *Classical SVM* (Boser et al. 1992) requires the constraint $y_j(w^\top \xi_j + b) \geq 0$ to hold for all j ("hard margin") and, as a result, may have no solution.
- *C-SVM* (Cortes and Vapnik 1995), also called SVM with "soft margin," allows the constraint $y_j(w^\top \xi_j + b) \geq 0$ to be violated for some j and has an additional parameter C that controls the extent to which the constraint is violated.
- *ν-SVM* (Schölkopf et al. 2000) is similar to the SVM with "soft margin," but uses parameter ν instead of C that controls the percentage of constraint violations (losses). It is equivalent to minimizing the weighted sum of $\|w\|^2$ and CVaR_ν of the loss function $f(w, b) = -y_j(w^\top \xi_j + b)$ (Crisp and Burges 2000; Takeda and Sugiyama 2008).
- *Extended ν-SVM*, also known as Eν-SVM (Perez-Cruz et al. 2003), minimizes CVaR_ν of the loss function $f(w, b)$ subject to a constraint on $\|w\|$. It is reported to outperform ν-SVM (Perez-Cruz et al. 2003).
- *Robust ν-SVM* (Wang 2012) is another extension of ν-SVM, in which training data may assume different empirical distributions.

This case study solves three SVM problems: ν-*SVM*, Eν-*SVM*, and robust ν-SVM with either VaR or CVaR regularization terms; see Tsyurmasto et al. (2014) for VaR-SVM.

References

Boser, B.E., Guyon, I.M., Vapnik, V.N. (1992): A training algorithm for optimal margin classifiers. In 5th Annual ACM Workshop on COLT, D. Haussler ed., Pittsburgh, PA, ACM Press, 144–152

Crisp, D.J., Burges, C.J.C. (2000): A geometric interpretation of ν-SVM classifiers. Neural Information Processing Systems **12**, 244–250

Cortes, C., Vapnik, V. (1995): Support-vector networks, Machine Learning **20**, 273–297

Perez-Cruz, F., Weston, J., Hermann, D.J.L., Schölkopf, B. (2003): Extension of the ν-SVM range for classification. Advances in Learning Theory: Methods, Models and Applications **190**, 179–196

Schölkopf, B., Smola, A., Williamson, R., Bartlett, P. (2000): New support vector algorithms, Neural Computation **12**, 1207–1245

Takeda, A., Sugiyama, M. (2008): ν-Support vector machine as conditional value-at-risk minimization. In: Proceedings of the 25th international conference on Machine learning, Morgan Kaufmann, Montreal, Canada, 1056–1063

Wang, Y. (2012): Robust ν-Support vector machine based on worst-case conditional value-at-risk minimization. Optimization Methods and Software **27**(6), 1025–1038

Tsyurmasto, P., Zabarankin, M., Uryasev, S. (2014): Value-at-Risk support vector machine: stability to outliers. Journal of Combinatorial Optimization, to appear

Notation	Description
J	Number of observations (interpreted as scenarios having equal probabilities)
$j \in \{1, \ldots, J\}$	Index of observations
(ξ_j, y_j)	Data of observation j
$\xi_j \in \mathbb{R}^n$	Vector of features in observation j
$y_j \in \{-1, 1\}$	Category label in observation j
$x = (w, b) \in \mathbb{R}^n \times \mathbb{R}$	Vector of decision variables
$L_j(x) = -y_j(w^\top \xi_j + b)$	Loss function for observation j
$L(x)$	Loss function treated as a discretely distributed random variable assuming values $L_1(x), \ldots, L_J(x)$ with equal probabilities of $1/J$
M	Number of data subsets
J_m	Number of observations in data subset m
$L^{(m)}(x)$	Loss function corresponding to data subset m

Main Functions

- **var_risk**$_\alpha(L(x))$ is *VaR for loss* defined by

$$\textbf{var\_risk}_\alpha(L(x)) = \arg\min_{c\in\mathbb{R}} \left(c + \frac{1}{1-\alpha} \sum_{j=1}^{J} p_j \max\left\{L_j(x) - c, 0\right\} \right),$$

  where $p_j = 1/J,\ j = 1, \ldots, J$.
- **cvar_risk**$_\alpha(L(x))$ is *CVaR for loss* [CVaR_α^-, see (8.6.1)] defined by

$$\textbf{cvar\_risk}_\alpha(L(x)) = \min_{c\in\mathbb{R}} \left(c + \frac{1}{1-\alpha} \sum_{j=1}^{J} p_j \max\left\{L_j(x) - c, 0\right\} \right),$$

  where $p_j = 1/J,\ j = 1, \ldots, J$.
- **max_var_risk**$_\alpha(L^{(1)}(x), \ldots, L^{(M)}(x))$ and **max_cvar_risk**$_\alpha(L^{(1)}(x), \ldots, L^{(M)}(x))$ are *maximum of VaRs* and *maximum of CVaRs*, respectively, over data subsets $m = 1, \ldots, M$ and are defined by

$$\textbf{max\_var\_risk}_\alpha(L^{(1)}(x), \ldots, L^{(M)}(x)) = \max_{m=1,\ldots,M} \textbf{var\_risk}_\alpha(L^{(m)}(x)),$$

$$\textbf{max\_cvar\_risk}_\alpha(L^{(1)}(x), \ldots, L^{(M)}(x)) = \max_{m=1,\ldots,M} \textbf{cvar\_risk}_\alpha(L^{(m)}(x)).$$

9.16.1 Problem 1a: ν-SVM with CVaR

Problem 9.35 (ν**-SVM with CVaR).** Minimize the weighted sum of the quadratic term $\|w\|^2$ and CVaR of the loss function $L(x)$:

$$\min_{x=(w,b)} \left(\frac{1}{2} w^\top w + \nu\, \textbf{cvar\_risk}_\alpha(L(x)) \right), \tag{9.16.1}$$

where $\nu = 1 - \alpha$.

Portfolio Safeguard Formulation

problem: problem_nu_svm, type = **minimize**
objective: objective_svm
0.5*quadratic_matrix_quadratic(**matrix_quadratic**)
0.5*cvar_risk_1(**0.5,matrix_prior_scenarios**)
box_of_variables: **upperbounds = 1000.0**, **lowerbounds = −1000.0**
solver: VAN, precision = 6, stages = 6

Problem Datasets

Data	http://www.ise.ufl.edu/uryasev/files/2012/07/data_problem_nu_svm.zip
Solution	http://www.ise.ufl.edu/uryasev/files/2012/07/solution_problem_nu_svm.zip
# of variables	24
# of scenarios	1,000
α	2
Objective value	0
Running time	0.29 s (PC 2.66 GHz)

9.16.2 Problem 1b: VaR-SVM

Problem 9.36 (VaR-SVM). Minimize the weighted sum of the quadratic term $\|w\|^2$ and VaR of the loss function $L(x)$:

$$\min_{x=(w,b)} \left(\frac{1}{2} w^\top w + \nu\, \textbf{var\_risk}_\alpha (L(x)) \right), \tag{9.16.2}$$

where, in general, α and ν are not related.

Portfolio Safeguard Formulation

problem: problem_nu_svm, type = **minimize**
objective: objective_svm
0.5*quadratic_matrix_quadratic(**matrix_quadratic**)
0.5*var_risk_1(**0.5,matrix_prior_scenarios**)
box_of_variables: **upperbounds = 1000.0**, **lowerbounds = −1000.0**
solver: VAN, precision = 6, stages = 6

Problem Datasets

Data	http://www.ise.ufl.edu/uryasev/files/2012/07/data_problem_var_nu_svm.zip
Solution	http://www.ise.ufl.edu/uryasev/files/2012/07/solution_problem_var_nu_svm.zip
# of variables	24
# of scenarios	1,000
α	0.5
Objective value	−7.06555e+03
Running time	1.54 s (PC 2.66 GHz)

9.16.3 Problem 2a: Extended ν-SVM with CVaR

Problem 9.37 (Eν-SVM with CVaR). Minimize CVaR of the loss function $L(x)$:

$$\min_{x=(w,b)} \textbf{cvar\_risk}_\alpha(L(x)) \tag{9.16.3}$$

subject to a constraint on the norm $\|w\|$:

$$w^\top w = 1. \tag{9.16.4}$$

Portfolio Safeguard Formulation

problem: problem_extended_nu_svm, type = **minimize**
objective: objective_svm
cvar_risk_1(**0.55,matrix_prior_scenarios**)
constraint: constraint_unity, **lower_bound = 1**, **upper_bound = 1**
quadratic_matrix_quadratic(**matrix_quadratic**)
box_of_variables: **upperbounds = 1000.0**, **lowerbounds = −1000.0**
solver: VAN, precision = 6, stages = 30

Problem Datasets

Data	http://www.ise.ufl.edu/uryasev/files/2012/07/data_problem_extended_nu_svm.zip
Solution	http://www.ise.ufl.edu/uryasev/files/2012/07/solution_problem_extended_nu_svm.zip
# of variables	24
# of scenarios	1,000
α	0.55
Objective value	0.261460
Running time	0.34 s (PC 2.66 GHz)

9.16.4 Problem 2b: Extended VaR-SVM

Problem 9.38 (Extended VaR-SVM). Minimize VaR of the loss function $L(x)$:

$$\min_{x=(w,b)} \mathbf{var\_risk}_\alpha(L(x)) \tag{9.16.5}$$

subject to a constraint on the norm $\|w\|$:

$$w^\top w = 1. \tag{9.16.6}$$

Portfolio Safeguard Formulation

problem: problem_extended_var_nu_svm, type = **minimize**
objective: objective_svm
var_risk_1(**0.25,matrix_prior_scenarios)**
constraint: constraint_unity, **lower_bound = 1**, **upper_bound = 1**
quadratic_matrix_quadratic(**matrix_quadratic)**
box_of_variables: **upperbounds = 1000.0**, **lowerbounds = −1000.0**
solver: VAN, precision = 6, stages = 30

Problem Datasets

Data	http://www.ise.ufl.edu/uryasev/files/2012/07/data_problem_extended_var_nu_svm.zip
Solution	http://www.ise.ufl.edu/uryasev/files/2012/07/solution_problem_extended_var_nu_svm.zip
# of variables	24
# of scenarios	1,000
α	0.25
Objective value	−1.05375e+04
Running time	2.4 s (PC 2.66 GHz)

9.16.5 Problem 3a: Robust ν-SVM with CVaR

Problem 9.39 (Robust ν-SVM with CVaR). Minimize the weighted sum of the quadratic term $\|w\|^2$ and the maximum of CVaRs of the loss function $L(x)$ over M data subsets:

$$\min_{x=(w,b)} \left(\frac{1}{2} w^{\top} w + \nu\, \mathbf{max\_cvar\_risk}_{\alpha}(L^{(1)}(x), \ldots, L^{(M)}(x)) \right), \tag{9.16.7}$$

where $\nu = 1 - \alpha$.

Portfolio Safeguard Formulation

problem: problem_robust_svm, type = **minimize**
objective: objective_svm
quadratic_matrix_quadratic(**matrix_quadratic**)
max_cvar_risk_1(0.5,matrix_prior_scenarios1,. . .,matrix_prior_scenarios5)
box_of_variables: **upperbounds = 1000.0, lowerbounds = −1000.0**
solver: VAN, precision = 6, stages = 6

Problem Datasets

Data	http://www.ise.ufl.edu/uryasev/files/2012/06/data_problem_robust_svm.zip
Solution	http://www.ise.ufl.edu/uryasev/files/2012/06/solution_problem_robust_svm.zip
# of variables	24
# of scenarios	134
α	0.5
Objective value	8.08515e–07
Running time	0.11 s (PC 2.66 GHz)

9.16.6 Problem 3b: Robust VaR-SVM

Problem 9.40 (Robust VaR-SVM). Minimize the weighted sum of the quadratic term $\|w\|^2$ and the maximum of VaRs of the loss function $L(x)$ over M data subsets:

$$\min_{x=(w,b)} \left(\frac{1}{2} w^\top w + \nu \, \mathbf{max\_var\_risk}_\alpha (L^{(1)}(x), \ldots, L^{(M)}(x)) \right), \tag{9.16.8}$$

where, in general, α and ν are not related.

Portfolio Safeguard Formulation

problem: problem_robust_var_svm, type = **minimize**
objective: objective_svm
0.5*quadratic_matrix_quadratic(**matrix_quadratic**)
0.5*max_var_risk_1(**0.45,matrix_prior_scenarios1,. . .,matrix_prior_scenarios5)**
box_of_variables: upperbounds = 1000.0, lowerbounds = −1000.0
solver: VAN, precision = 6, stages = 30

Problem Datasets

Data	http://www.ise.ufl.edu/uryasev/files/2012/06/data_problem_robust_svm.zip
Solution	http://www.ise.ufl.edu/uryasev/files/2012/06/solution_problem_robust_svm.zip
# of variables	24
# of scenarios	134
α	0.45
Objective value	−1,000
Running time	4.18 s (PC 2.66 GHz)

9.17 Case Study: Implied Copula CDO Pricing Model: Entropy Approach (Meansquare, Variance, Entropy)

Background

This case study estimates default distributions of a basket of credit default swaps (CDSs) based on prices of collateralized debt obligation (CDO) tranches. Two optimization problems are considered: the first minimizes deviation of tranche payoffs from perfect fit with respect to unknown default probabilities [the objective function also includes a smoothing term; see Hull and White (2006)], whereas the second maximizes the Shannon entropy of an unknown default distribution subject to *no-arbitrage constraints* and subject to constraints imposing a convex–concave–convex (CCC) shape (bell-shape) on the probability distribution; see Examples 5.4 and 5.5 and Veremyev et al. (2013).

References

Hull, J., White, A. (2006): Valuing credit derivatives using an implied copula approach. The Journal of Derivatives **14**(2), 8–28

Veremyev, A., Tsyurmasto, P., Uryasev, S., Rockafellar, R.T. (2013): Calibrating Probability Distributions with Convex-Concave-Convex Functions: Application to CDO Pricing. Computational Management Science, to appear

Notation	Description
J	Total number of tranches
$j \in \{1, \ldots, J\}$	Index of tranches
I	Total number of hazard rate scenarios
$i \in \{1, \ldots, I\}$	Index of hazard rate scenarios
$d_1, \ldots, d_I$	Hazard rate scenarios
p_i	Probability of hazard rate scenario i
a_{ij}	Net payoff of tranche j for hazard rate scenario i
$\underline{a}_{ij}$	Expected net payoff of tranche j for hazard rate scenario i for ask price
$\overline{a}_{ij}$	Expected net payoff of tranche j for hazard rate scenario i for bid price
$w_l, w_r \in \{1, \ldots, I\}$	Indices of the left and right inflection points in CCC distribution

Main Functions

- Shannon entropy is implemented through the PSG function

$$\textbf{entropyr}(p) = \sum_{i=1}^{I} p_i \ln\left(\frac{p_i}{\theta_i}\right) \qquad \text{for } p_i \geq 0 \text{ and } \theta_i = 1/I, \quad i = 1, \ldots, I.$$

- Deviation term $D(p)$ and smoothing term $S(p)$ are introduced by

$$D(p) = \sum_{j=1}^{J}\left(\sum_{i=1}^{I} p_i a_{ij}\right)$$

and

$$S(p) = c\sum_{i=2}^{I-1}\left(\frac{p_{i+1} - 2p_i + p_{i-1}}{0.5(d_{i+1} - d_{i-1})}\right),$$

respectively, where c is a trade-off coefficient.
- $D(p) + S(p)$ can be expressed either as the *meansquare* function:

$$\textbf{meansquare}(p, B) \equiv \sum_{j=1}^{J}\left(\sum_{i=1}^{I} p_i b_{ji}\right),$$

with some matrix B whose entries b_{ji} depend only on a_{ij} and d_i, or as the *variance* function:

$$\textbf{variance}(p, B) \equiv pBB^\top p^\top,$$

where $B^\top$ is the transpose of B.

9.17.1 Problem 1a: problem_hull_meansquare

Problem 9.41. Minimize the sum of the deviation term and smoothing term:

$$\min_p \textbf{meansquare}(p, B) \tag{9.17.1}$$

subject to probability distribution constraints

$$\sum_{i=1}^{I} p_i = 1, \qquad p_i \geq 0, \quad i = 1, \ldots, I. \tag{9.17.2}$$

Portfolio Safeguard Formulation

problem: problem_hull_meansquare, type = **minimize**
objective: objective_hull_meansquare
meansquare_function1(**matrix_meansquare1)**
0.0189*meansquare_function2(**matrix_meansquare2)**
constraint: constraint_aeq, **lower_bound = 1.00**, **upper_bound = 1.00**
linearmulti_aeq(**matrix_aeq)**
box_of_variables: **lowerbounds = point_lb**, **upperbounds = point_ub**
solver: VAN, precision = 9, stages = 6

Problem Datasets: Dataset1

Data	http://www.ise.ufl.edu/uryasev/files/2012/06/data_problem_Hull_meansquare.zip
Solution	http://www.ise.ufl.edu/uryasev/files/2012/06/solution_problem_Hull_meansquare.zip
# of variables	1,000
# of scenarios	6
Objective value	8.93762e–10
Running time	9.04 s (PC 2.66 GHz)

9.17.2 Problem 1b: problem_hull_variance

Problem 9.42. Minimize the sum of the deviation term and smoothing term:

$$\min_{p} \mathbf{variance}(p, B) \tag{9.17.3}$$

subject to probability distribution constraints

$$\sum_{i=1}^{I} p_i = 1, \qquad p_i \geq 0, \quad i = 1, \ldots, I. \tag{9.17.4}$$

Portfolio Safeguard Formulation

problem: problem_hull_variance, type = **minimize**
objective: objective_h, linearize = 1
variance_h(**matrix_h**)
constraint: constraint_aeq, **lower_bound = 1.00, upper_bound = 1.00**
linearmulti_aeq(**matrix_aeq**)
box_of_variables: **lowerbounds = point_lb, upperbounds = point_ub**
solver: VAN, precision = 6, stages = 6

Problem Datasets: Dataset1

Data	http://www.ise.ufl.edu/uryasev/files/2012/06/data_problem_Hull_variance.zip
Solution	http://www.ise.ufl.edu/uryasev/files/2012/06/solution_problem_Hull_variance.zip
# of variables	1,000
# of scenarios	1,000
Objective value	2.54403e–07
Running time	2 s (PC 2.66 GHz)

9.17.3 Problem 2: problem_CCC

Problem 9.43. Minimize the negative Shannon entropy

$$\min_{p} \mathbf{entropyr}(p) \tag{9.17.5}$$

subject to *no-arbitrage constraints*

$$\begin{aligned} &\sum_{i=1}^{I} \underline{a}_{ij} p_i \le 0, \quad j = 1, \dots, J, \\ &\sum_{i=1}^{I} \overline{a}_{ij} p_i \ge 0, \quad j = 1, \dots, J, \end{aligned} \tag{9.17.6}$$

"bell-shape" constraints

$$\begin{aligned} &\frac{p_{i-1} + p_{i+1}}{2} \ge p_i, \quad i = 2, \dots, w_l - 1 \quad &\text{(left slope is convex)} \\ &\frac{p_{i-1} + p_{i+1}}{2} \le p_i, \quad i = w_l + 1, \dots, w_r - 1 \quad &\text{(hump is concave)} \\ &\frac{p_{i-1} + p_{i+1}}{2} \ge p_i, \quad i = w_r + 1, \dots, n - 1 \quad &\text{(right slope is convex),} \end{aligned} \tag{9.17.7}$$

and subject to probability distribution constraints

$$\sum_{i=1}^{I} p_i = 1, \qquad p_i \ge 0, \quad i = 1, \dots, I. \tag{9.17.8}$$

Portfolio Safeguard Formulation

problem: problem_CCC, type = **minimize**
objective: objective_h, linearize = 1
entropyr_h(**matrix_h**)
constraint: constraint_a, **lower_bound = vector_bl**, **upper_bound = vector_b**
linearmulti_aeq(**matrix_aeq**)
box_of_variables: **lowerbounds = 0.0**
solver: VAN, precision = 5, stages = 6

Problem Datasets: Dataset1

Data	http://www.ise.ufl.edu/uryasev/files/2012/06/data_problem_CCC.zip
Solution	http://www.ise.ufl.edu/uryasev/files/2012/06/solution_problem_CCC.zip
# of variables	100
# of scenarios	112
Objective value	−3.47386
Running time	1.03 s (PC 2.66 GHz)

9.18 Case Study: Logistic Regression and Regularized Logistic Regression for Estimating the Probability of Cesarean Section (logexp_sum, polynom_abs)

Background

This case study solves regular ("plain vanilla") logistic regression and a regularized logistic regression to predict the probability of the cesarean section (CS) in a population as a function of six demographic factors: age, height, weight, maternal weight gain, gestational age, and birth weight. The "plain vanilla" logistic regression maximizes the log-likelihood function as in Chen et al. (2004) and is implemented with the PSG function **logexp_sum**, whereas the regularized logistic regression maximizes the difference of the log-likelihood function and a regularization term, which is implemented by the PSG function **polynom_abs**. The regularization term improves the out-of-sample performance of the regression model (Shi et al. 2008). The "plain vanilla" logistic regression and the regularized logistic regression will be referred to as first model and second model, respectively.

The case study makes four passes for the available dataset (including values of thirty nine factors and resulting outcomes). In each pass, both models are calibrated on 3/4 of the data (in-sample), and then they are tested on the remaining 1/4 of the data (out-of-sample). The aggregated out-of-sample performance of each model is measured by the average value of the log-likelihood function over the four artificially created out-of-sample parts of the data. In the second model, the coefficients of the function **polynom_abs** optimize out-of-sample performance and are obtained iteratively with a steepest descent algorithm.

References

Chen, G., Uryasev, S., Young, T.K. (2004): On the prediction of the cesarean delivery risk in a large private practice. American Journal of Obstetrics and Gynecology **191**, 617–625

Shi, W., Wahba, G., Wright, S., Lee, K., Klein, R., Klein, B. (2008): LASSO-patternsearch algorithm with application to ophthalmology and genomic data. Statistical Interface **1**(1), 137–153

Notation	Description
I	Number of factors
$i \in \{1, \dots, I\}$	Factor index
J	Number of scenarios
$j \in \{1, \dots, J\}$	Scenario index
$x = (x_0, x_1, \dots, x_I)$	Vector of decision variables
θ_{ij}	Value of factor i in scenario j
$\theta_{j0} \in \{0, 1\}$	Benchmark value in scenario j

Main Functions

- In the logistic regression, the log-likelihood function is implemented by the PSG function

$$\mathbf{logexp\_sum}(x) = \sum_{j=1}^{J}\left((1-\theta_{j0})\left[x_0 + \sum_{i=1}^{I}\theta_{ij}x_i\right] - \ln\left(1+\exp\left[x_0 + \sum_{i=1}^{I}\theta_{ij}x_i\right]\right)\right).$$

- The polynomial absolute regularization function ($\mathscr{L}_1$-vector norm) is defined by

$$\mathbf{polynom\_abs}(x) = \sum_{i=0}^{I}\eta_i\,|x_i|,$$

where the coefficients $\eta_0, \ldots, \eta_I$ will be optimized for out-of-sample performance.

9.18.1 Problem 1: Logistic Regression

Problem 9.44. Maximize the log-likelihood function

$$\max_x \mathbf{logexp\_sum}(x) \tag{9.18.1}$$

subject to box constraints on the decision vector

$$l_i \leq x_i \leq u_i, \qquad i = 1, \ldots, I. \tag{9.18.2}$$

Portfolio Safeguard Formulation

problem: problem_CSection_LOGEXPSUM_POLYNOMABS_MaxLikelihood, type = **maximize**
objective: objective_max_likelihood
logexp_sum_matrix_scenarios(**matrix_insample_1**)
box_of_variables: **lowerbounds = point_lowerbounds**, **upperbounds = point_upperbounds**
solver: CAR, precision = 4, stages = 6

Problem Datasets: Dataset1

Data	http://www.ise.ufl.edu/uryasev/files/2012/08/data_problem_CSection_LOGEXPSUM_POLYNOMABS_MaxLikelihood.zip
Solution	http://www.ise.ufl.edu/uryasev/files/2012/08/solution_problem_CSection_LOGEXPSUM_POLYNOMABS_MaxLikelihood.zip
# of variables	6
# of scenarios	16,921
Objective value	−0.495793
Running time	0.72 s (PC 2.66 GHz)

9.18.2 Problem 2: Regularized Logistic Regression

Problem 9.45. Maximize the regularized log-likelihood function

$$\max_x \left(\textbf{logexp\_sum}(x) - \textbf{polynom\_abs}(x)\right) \tag{9.18.3}$$

subject to box constraints on the decision vector

$$l_i \leq x_i \leq u_i, \qquad i = 1, \ldots, I. \tag{9.18.4}$$

Portfolio Safeguard Formulation

problem: problem_CSection_LOGEXPSUM_POLYNOMAMS_RegLikelihood,
type = **maximize**
objective: objective_regularized_likelihood
logexp_sum_matrix_scenarios(**matrix_insample_1)**
−**polynom_abs**_matrix_coefficients(**matrix_coefficients**)
box_of_variables: **lowerbounds = point_lowerbounds**,
upperbounds = point_upperbounds
solver: CAR, precision = 4, stages = 6

Problem Datasets: Dataset1

Data	http://www.ise.ufl.edu/uryasev/files/2012/08/data_problem_CSection_LOGEXPSUM_POLYNOMAMS_RegLikelihood.zip
Solution	http://www.ise.ufl.edu/uryasev/files/2012/08/solution_problem_CSection_LOGEXPSUM_POLYNOMAMS_RegLikelihood.zip
# of variables	6
# of scenarios	16,921
Objective value	−0.498204
Running time	0.62 s (PC 2.66 GHz)

9.19 Case Study: Portfolio Optimization with Drawdown Constraints on a Single Sample Path (cdar_dev, drawdown_dev_max, drawdown_dev_avg)

Background

This case study solves a portfolio optimization problem subject to a constraint on conditional drawdown-at-risk (CDaR) on a single sample path. Portfolio drawdown is a nonnegative function of time, which at each time moment is the drop of portfolio value from its current maximum. Negative drawdown is referred to as "underwater curve." Given a confidence level $\alpha \in [0, 1]$ and a sample path, CDaR is the mean of the worst $(1-\alpha)100\%$ drawdowns. It is often used in active portfolio management; see Chekhlov et al. (2003; 2005). Average and maximum drawdowns are limiting cases of CDaR: they correspond to $\alpha = 0$ and $\alpha = 1$, respectively. The optimization problem maximizes the annualized portfolio rate of return subject to constraints on "single sample path" CDaR for various α (including $\alpha = 0$ and $\alpha = 1$). An extension of this case study to multiple sample paths is discussed in Sect. 9.20.

References

Chekhlov, A., Uryasev S., Zabarankin, M. (2003): Portfolio optimization with drawdown constraints. In Asset and Liability Management Tools, B. Scherer ed., Risk Books, London, 263–278

Chekhlov, A., Uryasev S., Zabarankin, M. (2005): Drawdown measure in portfolio optimization, International Journal of Theoretical and Applied Finance **8**(1), 13–58

Notation	Description
I	Number of instruments in the portfolio
$i \in \{1, \ldots, I\}$	Index of instruments in the portfolio
J	Number of scenarios (time intervals)
$j \in \{1, \ldots, J\}$	Index of scenarios (time intervals), where scenario j corresponds to time interval j
$p_j = 1/J$	Probability of scenario j; $j = 1, \ldots, J$
N	Number of years in time period $[0, J]$
$x = (x_1, \ldots, x_I)$	Vector of portfolio weights
θ_{ji}	Rate of return of instrument i at time moment j
$x_{\min}$ and $x_{\max}$	Lower and upper bounds on portfolio weights, respectively
$\alpha \in [0, 1]$	Confidence level in portfolio CDaR
ν	Bound on portfolio CDaR

Main Functions

- The portfolio rate of return at time moment j is given by

$$G_j(x) = \sum_{i=1}^{I} \theta_{ji}\, x_i, \qquad j = 1, \ldots, J.$$

  Let $G(x, \theta) = \{G_1(x), \ldots, G_J(x)\}$.
- The portfolio uncompounded cumulative rate of return at time moment j is defined as

$$w_j(x) = \sum_{k=1}^{j} \sum_{i=1}^{I} \theta_{ki}\, x_i, \qquad j = 1, \ldots, J.$$

- The annualized rate of return of instrument i over time period $[0, J]$ is given by

$$\overline{r}_i = \frac{1}{N} \sum_{j=1}^{J} \theta_{ji}, \qquad i = 1, \ldots, I.$$

- The annualized portfolio rate of return over time period $[0, J]$ is determined by

$$R(x) = \frac{1}{N} w_J(x) = \frac{1}{N} \sum_{j=1}^{J} \sum_{i=1}^{I} \theta_{ji}\, x_i = \sum_{i=1}^{I} \overline{r}_i\, x_i.$$

- Portfolio drawdown is a nonnegative function of time defined at time moment j by

$$d_j(x) = \max_{0 \le t \le j} \{w_t(x)\} - w_j(x), \qquad j = 1, \dots, J.$$

- Maximum and average portfolio drawdowns over time period $[0, J]$ are given by

$$\textbf{drawdown\_dev\_max}(G(x,\theta)) = \max_{1 \le j \le J} d_j(x)$$

and

$$\textbf{drawdown\_dev\_avg}(G(x,\theta)) = \frac{1}{J}\sum_{j=1}^{J} d_j(x),$$

respectively.
- Let $d(x)$ be viewed as a random variable assuming values $d_1(x), \dots, d_J(x)$ with equal probabilities $p_1 = \dots = p_J = 1/J$; then CDaR over time period $[0, J]$ with confidence level α is introduced by

$$\textbf{cdar\_dev}(\alpha, G(x,\theta)) = \textbf{cvar\_risk}_\alpha(d(x)),$$

where $\textbf{cvar\_risk}_\alpha$ is *CVaR for loss* [or CVaR_α^-, see (8.6.1)] defined by

$$\textbf{cvar\_risk}_\alpha(d(x)) = \min_{c \in \mathbb{R}} \left(c + \frac{1}{1-\alpha} \sum_{j=1}^{J} p_j \max\left\{d_j(x) - c, 0\right\} \right).$$

9.19.1 Problem 1: Constraint on Maximum Drawdown

Problem 9.46. Maximize the average annualized portfolio rate of return $R(x)$:

$$\max_x \sum_{i=1}^{I} \bar{r}_i x_i \tag{9.19.1}$$

subject to a constraint on the maximum drawdown:

$$\textbf{drawdown\_dev\_max}(G(x,\theta)) \le \nu \tag{9.19.2}$$

and subject to box constraints on portfolio weights:

$$x_{\min} \le x_i \le x_{\max}, \qquad i = 1, \dots, I. \tag{9.19.3}$$

Portfolio Safeguard Formulation

problem: problem_max_drawdown_0p170, type = **maximize**
objective: objective_annualized_return, linearize = 0
linear_average_returns(**matrix_annualized_returns**)
constraint: constraint_max_drawdown, **upper_bound = 0.17**
drawdown_dev_max_scenarios(**matrix_scenarios**)
box_of_variables: lowerbounds = point_lowerbounds,
upperbounds = point_upperbounds
solver: VAN, precision = 6

Problem Datasets: Dataset1

Data	http://www.ise.ufl.edu/uryasev/files/2011/12/data_problem_max_drawdown_0p170.zip
Solution	http://www.ise.ufl.edu/uryasev/files/2011/12/solution_problem_max_drawdown_0p170.zip
# of variables	32
# of scenarios	1,166
x_{min}	0.2
x_{max}	0.8
Objective value	0.809763
Running time	0.01 s (PC 2.66 GHz)

9.19.2 Problem 2: Constraint on Average Drawdown

Problem 9.47. Maximize the average annualized portfolio rate of return $R(x)$:

$$\max_x \sum_{i=1}^{I} \overline{r}_i x_i \tag{9.19.4}$$

subject to a constraint on the average drawdown:

$$\textbf{drawdown\_dev\_avg}(G(x,\theta)) \leq \nu \tag{9.19.5}$$

and subject to box constraints on portfolio weights:

$$x_{\min} \leq x_i \leq x_{\max}, \qquad i = 1, \ldots, I. \tag{9.19.6}$$

Portfolio Safeguard Formulation

problem: problem_average_drawdown_0p0307, type = **maximize**
objective: objective_annualized_return, linearize = 0
linear_average_returns(**matrix_annualized_returns**)
constraint: constraint_average_drawdown, **upper_bound = 0.0307**
drawdown_dev_avg_scenarios(**matrix_scenarios**)
box_of_variables: **lowerbounds = point_lowerbounds**,
upperbounds = point_upperbounds
solver: VAN, precision = 6

Problem Datasets: Dataset1

Data	http://www.ise.ufl.edu/uryasev/files/2011/12/data_problem_average_drawdown_0p0307.zip
Solution	http://www.ise.ufl.edu/uryasev/files/2011/12/solution_problem_average_drawdown_0p0307.zip
# of variables	32
# of scenarios	1,166
$x_{\min}$	0.2
$x_{\max}$	0.8
Objective value	0.763602
Running time	0.01 s (PC 2.66 GHz)

9.19.3 Problem 3: Constraint on CDaR

Problem 9.48. Maximize the average annualized portfolio rate of return $R(x)$:

$$\max_x \sum_{i=1}^{I} \overline{r}_i x_i \tag{9.19.7}$$

subject to a constraint on CDaR:

$$\mathbf{cdar\_dev}(\alpha, G(x,\theta)) \leq \nu \tag{9.19.8}$$

and subject to box constraints on portfolio weights:

$$x_{\min} \leq x_i \leq x_{\max}, \qquad i = 1, \ldots, I. \tag{9.19.9}$$

Portfolio Safeguard Formulation

problem: problem_CDaR_0p110, type = **maximize**
objective: objective_annualized_return, linearize = 0
linear_average_returns(**matrix_annualized_returns**)
constraint: constraint_CDaR, **upper_bound = 0.110**
cdar_dev_scenarios(**0.95,matrix_scenarios**)
box_of_variables: **lowerbounds = point_lowerbounds,**
upperbounds = point_upperbounds
solver: VAN, precision = 6

Problem Datasets: Dataset1

Data	http://www.ise.ufl.edu/uryasev/files/2011/12/data_problem_CDAR_0p110.zip
Solution	http://www.ise.ufl.edu/uryasev/files/2011/12/solution_problem_CDAR_0p110.zip
# of variables	32
# of scenarios	1,166
$x_{\min}$	0.2
$x_{\max}$	0.8
α	0.95
Objective value	0.754324
Running time	0.01 s (PC 2.66 GHz)

9.20 Case Study: Portfolio Optimization with Drawdown Constraints on Multiple Sample Paths (cdarmulti_dev, drawdownmulti_dev_max, drawdownmulti_dev_avg)

Background

This case study solves a portfolio optimization problem subject to a constraint on CDaR on multiple sample paths. For each sample path, portfolio drawdown is a nonnegative function of time, which at each time moment is the drop of portfolio value from its current maximum. Negative drawdown is often referred to as "underwater curve." Given a confidence level $\alpha \in [0, 1]$ and multiple sample paths, CDaR is defined as the mean of the largest $(1 - \alpha)100\%$ drawdowns taken simultaneously over time and over all sample paths. It is often used in active portfolio management; see Chekhlov et al. (2003; 2005). Average and maximum drawdowns are limiting cases of CDaR: they correspond to $\alpha = 0$ and $\alpha = 1$, respectively. The optimization problem maximizes the annualized portfolio rate of return subject to constraints on "multiple sample paths" CDaR for various α (including $\alpha = 0$ and $\alpha = 1$). In this case study, each problem is solved for 2 datasets: Dataset1 includes 31 variables and 11 sample paths (12,925 scenarios), whereas Dataset2 contains 18 variables and 180 sample paths (211, 680 scenarios).

References

Chekhlov, A., Uryasev S., Zabarankin, M. (2003): Portfolio optimization with drawdown constraints. In Asset and Liability Management Tools, B. Scherer ed., Risk Books, London, 263–278

Chekhlov, A., Uryasev S., Zabarankin, M. (2005): Drawdown measure in portfolio optimization. International Journal of Theoretical and Applied Finance **8**(1), 13–58

Notation	Description
I	Number of instruments in the portfolio
$i \in \{1, \ldots, I\}$	Index of instruments in the portfolio
K	Number of sample paths (every sample path is a joint trajectory of rates of return of underlying instruments)
$k \in \{1, \ldots, K\}$	Index of sample paths
J	Number of scenarios (time intervals)
$j \in \{1, \ldots, J\}$	Index of scenarios (time intervals) in each sample path (every scenario is equally probable with probability $p_{kj} = 1/(KJ)$)
N	Number of years in time period $[0, J]$
$x = (x_1, \ldots, x_I)$	Vector of weights of instruments in the portfolio
θ_{kji}	Rate of return of instrument i at time moment j in sample path k
$x_{\min}$ and $x_{\max}$	Lower and upper bounds on portfolio weights, respectively
$\alpha \in [0, 1]$	Confidence level in portfolio CDaR
ν	Bound on portfolio CDaR

Main Functions

- The portfolio rate of return at time moment j in sample path k is given by

$$G_{kj}(x) = \sum_{i=1}^{I} \theta_{kji}\, x_i, \qquad j = 1, \ldots, J, \quad k = 1, \ldots, K.$$

Let $G_k(x, \theta) = \{G_{k1}(x), \ldots, G_{kJ}(x)\}$.

- The portfolio uncompounded cumulative rate of return at time moment j in sample path k is defined as

$$w_{kj}(x) = \sum_{m=1}^{j} \sum_{i=1}^{I} \theta_{kmi}\, x_i, \qquad j = 1, \ldots, J, \quad k = 1, \ldots, K.$$

- The annualized rate of return of instrument i over time period $[0, J]$ and averaged over K sample paths is given by

$$\overline{r}_i = \frac{1}{NK} \sum_{j=1}^{J} \sum_{k=1}^{K} \theta_{kji}, \qquad i = 1, \ldots, I.$$

- The annualized portfolio rate of return over time period $[0, J]$ and averaged over K sample paths is determined by

$$R(x) = \frac{1}{NK} \sum_{k=1}^{K} w_{kJ}(x) = \frac{1}{NK} \sum_{j=1}^{J} \sum_{i=1}^{I} \sum_{k=1}^{K} \theta_{kji}\, x_i = \sum_{i=1}^{I} \overline{r}_i\, x_i.$$

- For sample path k, portfolio drawdown is a nonnegative function of time defined at time moment j by

$$d_{kj}(x) = \max_{0 \le t \le j} \{w_{kt}(x)\} - w_{kj}(x), \qquad j = 1, \ldots, J, \quad k = 1, \ldots, K.$$

- Average and maximum portfolio drawdowns over time period $[0, J]$ and over K sample paths are given by

$$\textbf{drawdownmulti\_dev\_max}(G_1(x,\theta), \ldots, G_K(x,\theta)) = \max_{\substack{1 \le j \le J \\ 1 \le k \le K}} \{d_{kj}(x)\}$$

and

$$\textbf{drawdownmulti\_dev\_avg}(G_1(x,\theta), \ldots, G_K(x,\theta)) = \frac{1}{JK} \sum_{k=1}^{K} \sum_{j=1}^{J} d_{kj}(x),$$

respectively.
- Let $d(x)$ be viewed as a random variable assuming values $d_{kj}(x)$ with equal probabilities $p_{kj} = 1/(KJ)$, $j = 1, \ldots, J$, $k = 1, \ldots, K$. Then CDaR over time period $[0, J]$ and over K sample paths with confidence level α is introduced by

$$\textbf{cdarmulti\_dev}(\alpha, G_1(x,\theta), \ldots, G_K(x,\theta)) = \textbf{cvar\_risk}_\alpha(d(x)),$$

where **cvar_risk**$_\alpha$ is *CVaR for loss* [or CVaR_α^-, see (8.6.1)] defined by

$$\textbf{cvar\_risk}_\alpha(d(x)) = \min_{c\in\mathbb{R}}\left(c + \frac{1}{1-\alpha}\sum_{k=1}^{K}\sum_{j=1}^{J} p_{kj}\max\left\{d_{kj}(x) - c, 0\right\}\right).$$

9.20.1 Problem 1: Constraint on Maximum Drawdown

Problem 9.49. Maximize the average annualized portfolio rate of return $R(x)$:

$$\max_x \sum_{i=1}^{I} \overline{r}_i x_i \tag{9.20.1}$$

subject to a constraint on the maximum drawdown:

$$\textbf{drawdownmulti\_dev\_max}(G_1(x,\theta),\ldots,G_K(x,\theta)) \le \nu \tag{9.20.2}$$

and subject to box constraints on portfolio weights:

$$x_{\min} \le x_i \le x_{\max}, \qquad i = 1,\ldots,I. \tag{9.20.3}$$

Portfolio Safeguard Formulation: Dataset1

problem: problem_max_drawdown_0p08, type = **maximize**
objective: objective_average_final_return, linearize = 0
linear_average_returns(**matrix_average_annualized_returns**)
constraint: constraint_max_drawdown, **upper_bound = 0.08**
drawdownmulti_dev_max_constraint(**matrix_1,. . .,matrix_11**)
box_of_variables: **lowerbounds = point_lowerbounds,**
upperbounds = point_upperbounds
solver: VAN, precision = 6

Problem Datasets: Dataset1

Data	http://www.ise.ufl.edu/uryasev/files/2011/12/data_problem_cdar_drawdown_short.zip
Solution	http://www.ise.ufl.edu/uryasev/files/2011/12/solution_problem_max_drawdown_0p08_short.zip

# of variables	31
# of scenarios	12,925
$x_{\min}$	0.2
$x_{\max}$	0.8
Objective value	0.572829
Running time	0.04 s (PC 2.66 GHz)

Portfolio Safeguard Formulation: Dataset2

problem: problem_max_drawdown_0p08, type = **maximize**
objective: objective_average_final_returns, linearize = 0
linear_average_returns(**matrix_average_final_uncompound)**
constraint: constraint_max_drawdown, **upper_bound = 0.08**, linearize = 1
drawdownmulti_dev_max_constraint_1(**matrix_1,. . .,matrix_180)**
box_of_variables: **lowerbounds = point_lowerbounds,**
upperbounds = point_upperbounds
solver: VAN, precision = 6

Problem Datasets: Dataset2

Data	http://www.ise.ufl.edu/uryasev/files/2011/12/data_problem_cdar_drawdown.zip
Solution	http://www.ise.ufl.edu/uryasev/files/2011/12/solution_problem_max_drawdown_0p08.zip
# of variables	18
# of scenarios	211,680
$x_{\min}$	0.2
$x_{\max}$	0.8
Objective value	0.248019
Running time	0.31 s (PC 2.66 GHz)

9.20.2 Problem 2: Constraint on Average Drawdown

Problem 9.50. Maximize the average annualized portfolio rate of return $R(x)$:

$$\max_x \sum_{i=1}^{I} \overline{r}_i x_i \tag{9.20.4}$$

subject to a constraint on the average drawdown:

$$\textbf{drawdownmulti\_dev\_avg}(G_1(x,\theta),\ldots,G_K(x,\theta)) \leq \nu \tag{9.20.5}$$

and subject to box constraints on portfolio weights:

$$x_{\min} \leq x_i \leq x_{\max}, \qquad i = 1,\ldots,I. \tag{9.20.6}$$

Portfolio Safeguard Formulation: Dataset1

problem: problem_average_drawdown_0p009, type = **maximize**
objective: objective_average_final_return, linearize = 0
linear_average_returns(**matrix_average_annualized_returns)**
constraint: constraint_average_drawdown, **upper_bound = 0.009**
drawdownmulti_dev_avg_constraint(**matrix_1,. . .,matrix_11)**
box_of_variables: **lowerbounds = point_lowerbounds,**
upperbounds = point_upperbounds
solver: VAN, precision = 6

Problem Datasets: Dataset1

Data	http://www.ise.ufl.edu/uryasev/files/2011/12/data_problem_cdar_drawdown_short.zip
Solution	http://www.ise.ufl.edu/uryasev/files/2011/12/solution_problem_average_drawdown_0p009_short.zip
# of variables	31
# of scenarios	12,925
$x_{\min}$	0.2
$x_{\max}$	0.8
Objective value	0.276401
Running time	0.38 s (PC 2.66 GHz)

Portfolio Safeguard Formulation: Dataset2

problem: problem_average_drawdown_0p009, type = **maximize**
objective: objective_average_final_returns, linearize = 0
linear_average_returns(**matrix_average_final_uncompound**)
constraint: constraint_average_drawdown, **upper_bound = 0.009**, linearize = 1
drawdownmulti_dev_avg_constraint_1(**matrix_1,. . .,matrix_180**)
box_of_variables: **lowerbounds = point_lowerbounds,**
upperbounds = point_upperbounds
solver: TANK, precision = 6

Problem Datasets: Dataset2

Data	http://www.ise.ufl.edu/uryasev/files/2011/12/data_problem_cdar_drawdown.zip
Solution	http://www.ise.ufl.edu/uryasev/files/2011/12/solution_problem_average_drawdown_0p009.zip
# of variables	18
# of scenarios	211,680
$x_{\min}$	0.2
$x_{\max}$	0.8
Objective value	0.190918
Running time	14.00 s (PC 2.66 GHz)

9.20.3 Problem 3: Constraint on CDaR

Problem 9.51. Maximize the average annualized portfolio rate of return $R(x)$:

$$\max_x \sum_{i=1}^{I} \overline{r}_i x_i \qquad (9.20.7)$$

subject to a constraint on CDaR:

$$\textbf{cdarmulti\_dev}(\alpha, G_1(x,\theta), \ldots, G_K(x,\theta)) \le \nu \qquad (9.20.8)$$

and subject to box constraints on portfolio weights:

$$x_{\min} \le x_i \le x_{\max}, \qquad i = 1, \ldots, I. \qquad (9.20.9)$$

Portfolio Safeguard Formulation: Dataset1

problem: problem_CDaR_0p03, type = **maximize**
objective: objective_average_final_return, linearize = 0
linear_average_returns(**matrix_average_annualized_returns**)
constraint: constraint_CDaR, **upper_bound = 0.03**
cdarmulti_dev_constraint(**0.8,matrix_1,. . .,matrix_11**)
box_of_variables: lowerbounds = point_lowerbounds,
upperbounds = point_upperbounds
solver: VAN, precision = 6

Problem Datasets: Dataset1

Data	http://www.ise.ufl.edu/uryasev/files/2011/12/data_problem_cdar_drawdown_short.zip
Solution	http://www.ise.ufl.edu/uryasev/files/2011/12/solution_problem_CDAR_0p03_short.zip
# of variables	31
# of scenarios	12,925
$x_{\min}$	0.2
$x_{\max}$	0.8
α	0.8
Objective value	0.384147
Running time	0.51 s (PC 2.66 GHz)

Portfolio Safeguard Formulation: Dataset2

problem: problem_CDaR_0p03, type = **maximize**
objective: objective_average_final_returns
linear_average_returns(**matrix_average_final_uncompound**)
constraint: constraint_CDaR, **upper_bound = 0.03**, linearize = 1
cdarmulti_dev_constraint_1(**0.8,matrix_1,. . .,matrix_180**)
box_of_variables: **lowerbounds = point_lowerbounds,**
upperbounds = point_upperbounds
solver: VAN, precision = 6

Problem Datasets: Dataset2

Data	http://www.ise.ufl.edu/uryasev/files/2011/12/data_problem_cdar_drawdown.zip
Solution	http://www.ise.ufl.edu/uryasev/files/2011/12/solution_problem_CDAR_0p03.zip
# of variables	18
# of scenarios	211,680
$x_{\min}$	0.2
$x_{\max}$	0.8
α	0.8
Objective value	0.247423
Running time	9.06 s (PC 2.66 GHz)

9.21 Case Study: Portfolio Optimization with Drawdown Constraints: Single Path Versus Multiple Paths (cdar_dev, cdarmulti_dev)

Background

This case study solves two portfolio optimization problems, which both maximize the annualized expected portfolio rate of return subject to a constraint on CDaR; see Chekhlov et al. (2003; 2005). The first problem defines the portfolio rate of return and CDaR based on multiple sample paths for portfolio instruments. In this case, given confidence level $\alpha \in [0, 1]$, CDaR is the mean of the largest $(1-\alpha)100\%$ drawdowns taken simultaneously over time and over all sample paths (PSG function **cdarmulti_dev**). The second problem aggregates all sample paths into the single one and defines the portfolio rate of return and CDaR based on the aggregated sample path. In this case, CDaR is the mean of the largest $(1-\alpha)100\%$ drawdowns in the single aggregated sample path (PSG function **cdar_dev**). Both problems are solved for "long" and "short" datasets: Dataset1 includes 180 sample paths, whereas Dataset2 contains only 11 sample paths.

References

Chekhlov, A., Uryasev S., Zabarankin, M. (2003): Portfolio optimization with drawdown constraints. In Asset and Liability Management Tools, B. Scherer ed., Risk Books, London, 263–278

Chekhlov, A., Uryasev S., Zabarankin, M. (2005): Drawdown measure in portfolio optimization. International Journal of Theoretical and Applied Finance **8**(1), 13–58

Notation	Description
I	Number of instruments in the portfolio
$i \in \{1, \ldots, I\}$	Index of instruments in the portfolio
K	Number of sample paths (every sample path is a joint trajectory of rates of return of underlying instruments)
$k \in \{1, \ldots, K\}$	Index of sample paths
J	Number of scenarios (time intervals)
$j \in \{1, \ldots, J\}$	Index of scenarios (time intervals) in each sample path (every scenario is equally probable with probability $p_{kj} = 1/(KJ)$)
$l \in \{1, \ldots, KJ\}$	Index of scenarios in the aggregated sample path formed by successively joining sample paths $1, \ldots, K$; in this case, every scenario is equally probable with probability $p_l = 1/(KJ)$
N	Number of years in time period $[0, J]$
$x = (x_1, \ldots, x_I)$	Vector of weights of instruments in the portfolio
θ_{kji}	Rate of return of instrument i at time moment j in sample path k
θ_{li}	Rate of return of instrument i for scenario j in the aggregated sample path ($\theta_{li} = \theta_{kji}$, $l = (k-1)J + j$, $j = 1, \ldots, J$, $k = 1, \ldots, K$)
$x_{\min}$ and $x_{\max}$	Lower and upper bounds on portfolio weights, respectively
$\alpha \in [0, 1]$	Confidence level in portfolio CDaR
ν	Bound on portfolio CDaR

Main Functions for Multiple Sample Paths

- The portfolio rate of return at time moment j in sample path k is given by

$$G_{kj}(x) = \sum_{i=1}^{I} \theta_{kji}\, x_i, \qquad j = 1, \ldots, J, \quad k = 1, \ldots, K.$$

  Let $G_k(x, \theta) = \{G_{k1}(x), \ldots, G_{kJ}(x)\}$.
- The portfolio uncompounded cumulative rate of return at time moment j in sample path k is defined as

$$w_{kj}(x) = \sum_{m=1}^{j} \sum_{i=1}^{I} \theta_{kmi}\, x_i, \qquad j = 1, \ldots, J, \quad k = 1, \ldots, K.$$

- The annualized rate of return of instrument i over time period $[0, J]$ and averaged over K sample paths is given by

$$\overline{r}_i = \frac{1}{NK} \sum_{j=1}^{J} \sum_{k=1}^{K} \theta_{kji}, \qquad i = 1, \ldots, I.$$

- The annualized portfolio rate of return over time period $[0, J]$ and averaged over K sample paths is determined by

$$R(x) = \frac{1}{NK} \sum_{k=1}^{K} w_{kJ}(x) = \frac{1}{NK} \sum_{j=1}^{J} \sum_{i=1}^{I} \sum_{k=1}^{K} \theta_{kji} x_i = \sum_{i=1}^{I} \overline{r}_i\, x_i.$$

- For sample path k, portfolio drawdown is a nonnegative function of time defined at time moment j by

$$d_{kj}(x) = \max_{0 \le t \le j} \{w_{kt}(x)\} - w_{kj}(x), \qquad j = 1, \ldots, J, \quad k = 1, \ldots, K.$$

- Let $d(x)$ be viewed as a random variable assuming values $d_{kj}(x)$ with equal probabilities $p_{kj} = 1/(KJ)$, $j = 1, \ldots, J$, $k = 1, \ldots, K$. Then CDaR over time period $[0, J]$ and over K sample paths with confidence level α is introduced by

$$\textbf{cdarmulti\_dev}(\alpha, G_1(x, \theta), \ldots, G_K(x, \theta)) = \textbf{cvar\_risk}_\alpha(d(x)),$$

where $\textbf{cvar\_risk}_\alpha$ is *CVaR for loss* [or CVaR_α^-, see (8.6.1)] defined by

$$\textbf{cvar\_risk}_\alpha(d(x)) = \min_{c \in \mathbb{R}} \left(c + \frac{1}{1-\alpha} \sum_{k=1}^{K} \sum_{j=1}^{J} p_{kj} \max\left\{d_{kj}(x) - c, 0\right\} \right).$$

Main Functions for the Aggregated Path

- The portfolio rate of return at time moment j in the aggregated sample path is given by

$$G_l(x) = \sum_{i=1}^{I} \theta_{li}\, x_i, \qquad l = 1, \ldots, KJ.$$

Let $G(x, \theta) = \{G_1(x), \ldots, G_{KJ}(x)\}$.

- The portfolio uncompounded cumulative rate of return at time moment l in the aggregated sample path is defined as

$$w_l(x) = \sum_{m=1}^{l} \sum_{i=1}^{I} \theta_{mi}\, x_i, \qquad l = 1, \ldots, KJ.$$

- The annualized rate of return of instrument i over time period $[0, KJ]$ is given by

$$\overline{r}_i = \frac{1}{KN} \sum_{l=1}^{KJ} \theta_{li}, \qquad i = 1, \ldots, I,$$

which coincides with $\overline{r}_i$ in the case of multiple sample paths.
- The annualized portfolio rate of return over time period $[0, KJ]$ is determined by

$$R(x) = \frac{1}{KN} w_{KJ}(x) = \frac{1}{KN} \sum_{l=1}^{KJ} \sum_{i=1}^{I} \theta_{li}\, x_i = \sum_{i=1}^{I} \overline{r}_i\, x_i,$$

which coincides with $R(x)$ in the case of multiple sample paths.
- For the aggregated sample path, portfolio drawdown is a nonnegative function of time defined at time moment l by

$$d_l(x) = \max_{0 \le t \le l} \{w_t(x)\} - w_l(x), \qquad l = 1, \ldots, KJ.$$

- Let $\tilde{d}(x)$ be viewed as a random variable assuming values $d_1(x), \ldots, d_{KJ}(x)$ with equal probabilities $p_1 = \ldots = p_{KJ} = 1/(KJ)$. Then CDaR for the aggregated sample path with confidence level α is introduced by

$$\textbf{cdar\_dev}(\alpha, G(x,\theta)) = \textbf{cvar\_risk}_\alpha \left(\tilde{d}(x)\right),$$

where $\textbf{cvar\_risk}_\alpha$ is *CVaR for loss* [or CVaR_α^-, see (8.6.1)] defined by

$$\textbf{cvar\_risk}_\alpha \left(\tilde{d}(x)\right) = \min_{c \in \mathbb{R}} \left(c + \frac{1}{1-\alpha} \sum_{j=1}^{KJ} p_j \max\left\{d_j(x) - c, 0\right\} \right).$$

9.21.1 Problem 1: Multiple Sample Paths

Problem 9.52. Maximize the average annualized portfolio rate of return $R(x)$:

$$\max_{x} \sum_{i=1}^{I} \overline{r}_i x_i \qquad (9.21.1)$$

subject to a constraint on CDaR:

$$\textbf{cdarmulti\_dev}(\alpha, G_1(x,\theta), \ldots, G_K(x,\theta)) \le \nu \qquad (9.21.2)$$

and subject to box constraints on portfolio weights:

$$x_{\min} \le x_i \le x_{\max}, \qquad i = 1, \ldots, I. \qquad (9.21.3)$$

Portfolio Safeguard Formulation: Dataset1

problem: problem_CDaR_multiple, type = **maximize**
objective: objective_average_final_return
linear_average_returns(**matrix_average_final_uncompound)**
constraint: constraint_CDaR_multiple, **upper_bound = 0.023**
cdarmulti_dev_constraint_multiple(**0.8,matrix_1,. . .,matrix_180)**
box_of_variables: lowerbounds = point_lowerbounds,
upperbounds = point_upperbounds
solver: VAN, precision = 9, stages = 6

Problem Datasets: Dataset1

Data	http://www.ise.ufl.edu/uryasev/files/2011/12/data_problem_cdar_dev_multiple_long.zip
Solution	http://www.ise.ufl.edu/uryasev/files/2011/12/solution_problem_cdar_multiple_long.zip
# of variables	18
# of scenarios	211,618
$x_{\min}$	0.2
$x_{\max}$	0.8
α	0.8
Objective value	0.182599
Running time	52.74 s (PC 2.66 GHz)

Portfolio Safeguard Formulation: Dataset2

problem: problem_CDaR_multiple, type = **maximize**
objective: objective_average_final_return
linear_average_returns(**matrix_average_annualized_returns**)
constraint: constraint_CDaR_multiple, **upper_bound = 0.023**
cdarmulti_dev_constraint_multiple(**0.8,matrix_1,. . .,matrix_11**)
box_of_variables: **lowerbounds = point_lowerbounds,**
upperbounds = point_upperbounds
solver: VAN, precision = 9, stages = 6

Problem Datasets: Dataset2

Data	http://www.ise.ufl.edu/uryasev/files/2011/12/data_problem_cdar_dev_multiple_short.zip
Solution	http://www.ise.ufl.edu/uryasev/files/2011/12/solution_problem_cdar_dev_multiple_short.zip
# of variables	30
# of scenarios	12,925
$x_{\min}$	0.2
$x_{\max}$	0.8
α	0.8
Objective value	0.240832
Running time	0.65 s (PC 2.66 GHz)

9.21.2 Problem 2: Aggregated Sample Path

Problem 9.53. Maximize the average annualized portfolio rate of return $R(x)$:

$$\max_x \sum_{i=1}^{I} \overline{r}_i x_i \tag{9.21.4}$$

subject to a constraint on CDaR:

$$\textbf{cdar\_dev}(\alpha, G(x,\theta)) \leq \nu \tag{9.21.5}$$

and subject to box constraints on portfolio weights:

$$x_{\min} \leq x_i \leq x_{\max}, \qquad i = 1, \ldots, I. \tag{9.21.6}$$

Portfolio Safeguard Formulation: DataSet1

problem: problem_CDaR_single, type = **maximize**
objective: objective_average_final_return
linear_average_returns(**matrix_average_final_uncompound)**
constraint: constraint_CDaR_single, **upper_bound = 0.023**
cdar_dev_constraint_single(**0.8,matrix_h**)
box_of_variables: **lowerbounds = point_lowerbounds,**
upperbounds = point_upperbounds
solver: VAN, precision = 9, stages = 6

Problem Datasets: Dataset1

Data	http://www.ise.ufl.edu/uryasev/files/2011/12/data_problem_cdar_dev_single_long.zip
Solution	http://www.ise.ufl.edu/uryasev/files/2011/12/solution_problem_cdar_single_long.zip
# of variables	18
# of scenarios	211,618
$x_{\min}$	0.2
$x_{\max}$	0.8
α	0.8
Objective value	0.179377
Running time	42.94 s (PC 2.66 GHz)

Portfolio Safeguard Formulation: DataSet2

problem: problem_CDaR_single, type = **maximize**
objective: objective_average_final_return
linear_average_returns(**matrix_average_annualized_returns**)
constraint: constraint_CDaR_single, **upper_bound = 0.023**
cdar_dev_constraint_single(**0.8,matrix_h**)
box_of_variables: **lowerbounds = point_lowerbounds,**
upperbounds = point_upperbounds
solver: VAN, precision = 9, stages = 6

Problem Datasets: Dataset2

Data	http://www.ise.ufl.edu/uryasev/files/2011/12/data_problem_cdar_dev_single_short.zip
Solution	http://www.ise.ufl.edu/uryasev/files/2011/12/solution_problem_cdar_dev_single_short.zip
# of variables	30
# of scenarios	12,925
$x_{\min}$	0.2
$x_{\max}$	0.8
α	0.8
Objective value	0.228868
Running time	0.5 s (PC 2.66 GHz)

References

[1] Acerbi, C.: Spectral measures of risk: a coherent representation of subjective risk aversion. J. Bank. Financ. **26**(7), 1487–1503 (2002)

[2] Artzner, P., Delbaen, F., Eber, J.M., Heath, D.: Coherent measures of risk. Math. Financ. **9**(3), 203–228 (1999)

[3] Bonami, P., Lejeune, M.A.: An exact solution approach for portfolio optimization problems under stochastic and integer constraints. Oper. Res. **57**(3), 650–670 (2009)

[4] Buckley, J.J.: Entropy principles in decision making under risk. Risk Anal. **5**(4), 303–313 (1979)

[5] Chang, C.C., Lin, C.J.: Training ν-support vector classifiers: theory and algorithms. Neural Comput. **13**, 2119–2147 (2001)

[6] Chekhlov, A., Uryasev, S., Zabarankin, M.: Portfolio Optimization with Drawdown Constraints, pp. 263–278. Risk Books, London (2003)

[7] Chekhlov, A., Uryasev, S., Zabarankin, M.: Drawdown measure in portfolio optimization. Int. J. Theor. Appl. Financ. **8**(1), 13–58 (2005)

[8] Cortes, C., Vapnik, V.: Support vector networks. Mach. Learn. **20**, 273–297 (1995)

[9] Costa, J., Hero, A., Vignat, C.: On Solutions to Multivariate Maximum-entropy Problems, vol. 2683, pp. 211–228. Springer, Berlin (2003)

[10] Cover, T.M., Thomas, J.A.: Elements of Information Theory, 2nd edn. Wiley, New York (2006)

[11] Cozzolino, J.M., Zahner, M.J.: The maximum-entropy distribution of the future market price of a stock. Oper. Res. **21**(6), 1200–1211 (1973)

[12] Crisp, D.J., Burges, C.J.C.: A geometric interpretation of ν-SVM classifiers. Neural Inf. Process. Syst. **12**, 244–250 (2000)

[13] Fölmer, H., Schied, A.: Stochastic Finance, 2nd edn. Walter de Gruyter GmbH & Co., Berlin (2004)

[14] Grauer, R.R.: Introduction to asset pricing theory and tests. In: Roll, R. (ed.) The International Library of Critical Writings in Financial Economics. Edward Elgar Publishing Inc., Cheltenham (2001)

[15] Grechuk, B., Zabarankin, M.: Inverse portfolio problem with mean-deviation model. Eur. J. Oper. Res. (2013, to appear)

[16] Grechuk, B., Molyboha, A., Zabarankin, M.: Maximum entropy principle with general deviation measures. Math. Oper. Res. **34**(2), 445–467 (2009)

[17] Grechuk, B., Molyboha, A., Zabarankin, M.: Chebyshev's inequalities with law invariant deviation measures. Probab. Eng. Informational Sci. **24**, 145–170 (2010)

[18] Hardy, G.E., Littlewood, J.E., Pólya, G.: Inequalities, 2nd edn. Cambridge University Press, New York (1952)

M. Zabarankin and S. Uryasev, *Statistical Decision Problems: Selected Concepts and Portfolio Safeguard Case Studies*, Springer Optimization and Its Applications 85, DOI 10.1007/978-1-4614-8471-4, © Springer Science+Business Media New York 2014

[19] Hastie, T., Tibshirani, R., Friedman, J.: The Elements of Statistical Learning: Data Mining, Inference, and Prediction, 2nd edn. Springer, New York (2008)

[20] Hull, J.C., White, A.D.: Valuing credit derivatives using an implied copula approach. J. Derivatives **14**(2), 8–28 (2006)

[21] Iscoe, I., Kreinin, A., Mausser, H., Romanko, A.: Portfolio credit-risk optimization. J. Bank. Financ. **36**(6), 1604–1615 (2012)

[22] Jaynes, E.T.: Information theory and statistical mechanics. Phys. Rev. **106**(4), 620–630 (1957)

[23] Jensen, J.L.: Surles fonctions convexes et les inégalités entre les valeurs moyennes. Acta Math. **30**(1), 175–193 (1906)

[24] Johnson, O., Vignat, C.: Some results concerning maximum Rényi entropy distributions. Annales de l'Institut Henri Poincare (B) Probab. Stat. **43**(3), 339–351 (2007)

[25] Kalinchenko, K., Uryasev, S., Rockafellar, R.T.: Calibrating risk preferences with generalized CAPM based on mixed CVaR deviation. J. Risk **15**(1), 45–70 (2012)

[26] Koenker, R., Bassett, G.: Regression quantiles. Econometrica **46**, 33–50 (1978)

[27] Kurdila, A., Zabarankin, M.: Convex Functional Analysis. Birkhauser, Switzerland (2005)

[28] Levy, H.: Stochastic dominance and expected utility: survey and analysis. Manag. Sci. **38**(4), 555–593 (1992)

[29] Markowitz, H.M.: Portfolio selection. J. Financ. **7**(1), 77–91 (1952)

[30] Markowitz, H.M.: Foundations of portfolio theory. J. Financ. **46**, 469–477 (1991)

[31] Mercer, J.: Functions of positive and negative type, and their connection with the theory of integral equations. Philos. Trans. Roy. Soc. London **209**(441–458), 415–446 (1909)

[32] Molyboha, A., Zabarankin, M.: Stochastic optimization of sensor placement for diver detection. Oper. Res. **60**(2), 292–312 (2012)

[33] Ogryczak, W., Ruszczyński, A.: On consistency of stochastic dominance and mean-semideviation models. Math. Program. **89**, 217–232 (2001)

[34] Ogryczak, W., Ruszczyński, A.: Dual stochastic dominance and related mean-risk models. SIAM J. Optim. **13**(1), 60–78 (2002)

[35] Perez-Cruz, F., Weston, J., Hermann, D.J.L., Schölkopf, B.: Extension of the ν-SVM range for classification. Adv. Learn. Theory Method. Models Appl. **190**, 179–196 (2003)

[36] Rockafellar, R.T.: Convex Analysis, Princeton Mathematics Series, vol. 28. Princeton University Press, Princeton (1970)

[37] Rockafellar, R.T.: Coherent approaches to risk in optimization under uncertainty. In: Gray, P. (ed.) Tutorials in Operations Research, pp. 38–61. INFORMS, Hanover (2007)

[38] Rockafellar, R.T., Royset, J.O.: On buffered failure probability in design and optimization of structures. Reliab. Eng. Syst. Saf. **95**, 499–510 (2011)

[39] Rockafellar, R.T., Uryasev, S.: Optimization of conditional value-at-risk. J. Risk **2**, 21–41 (2000)

[40] Rockafellar, R.T., Uryasev, S.: Conditional value-at-risk for general loss distributions. J. Bank. Financ. **26**(7), 1443–1471 (2002)

[41] Rockafellar, R.T., Uryasev, S., Zabarankin, M.: Deviation measures in risk analysis and optimization. Technical Report 2002–7. ISE Department, University of Florida, Gainesville, FL (2002)

[42] Rockafellar, R.T., Uryasev, S., Zabarankin, M.: Generalized deviations in risk analysis. Financ. Stoch. **10**(1), 51–74 (2006)

[43] Rockafellar, R.T., Uryasev, S., Zabarankin, M.: Master funds in portfolio analysis with general deviation measures. J. Bank. Financ. **30**(2), 743–778 (2006)

[44] Rockafellar, R.T., Uryasev, S., Zabarankin, M.: Optimality conditions in portfolio analysis with general deviation measures. Math. Program. **108**(2–3), 515–540 (2006)

[45] Rockafellar, R.T., Uryasev, S., Zabarankin, M.: Equilibrium with investors using a diversity of deviation measures. J. Bank. Financ. **31**(11), 3251–3268 (2007)

[46] Rockafellar, R.T., Uryasev, S., Zabarankin, M.: Risk tuning with generalized linear regression. Math. Oper. Res. **33**(3), 712–729 (2008)

[47] Roell, A.: Risk aversion in Quiggin and Yaari's rank-order model of choice under uncertainty. Econ. J. **97**(Issue Supplement: Conference papers), 143–159 (1987)

[48] Rousseeuw, P.J., Driessen, K.: Computing LTS regression for large data sets. Data Min. Knowl. Discov. **12**(1), 29–45 (2006)

[49] Rousseeuw, P., Leroy, A.: Robust Regression and Outlier Detection. Wiley, New York (1987)

[50] Roy, A.D.: Safety first and the holding of assets. Econometrica **20**(3), 431–449 (1952)

[51] Ruszczyński, A.: Nonlinear Optimization. Princeton University Press, Princeton (2006)

[52] Schölkopf, B., Smola, A., Williamson, R., Bartlett, P.: New support vector algorithms. Neural Comput. **12**, 1207–1245 (2000)

[53] Sharpe, W.F.: Capital asset prices: a theory of market equilibrium under conditions of risk. J. Financ. **19**, 425–442 (1964)

[54] Sharpe, W.F.: Capital asset prices with and without negative holdings. J. Financ. **46**, 489–509 (1991)

[55] Takeda, A., Sugiyama, M.: ν-support vector machine as conditional value-at-risk minimization. In: Proceedings of the 25th International Conference on Machine Learning (ICML 2008), pp. 1056–1063. Morgan Kaufmann, Montreal, Canada (2008)

[56] Thomas, M.U.: A generalized maximum entropy principle. Oper. Res. **27**(6), 1188–1196 (1979)

[57] Tobin, J.: Liquidity preference as behavior towards risk. Rev. Econ. Stud. **25**(2), 65–86 (1958)

[58] Tsyurmasto, P., Zabarankin, M., Uryasev, S.: Value-at-risk support vector machine: stability to outliers. J. Comb. Optim. (2014, to appear)

[59] Venables, W., Ripley, B.: Modern Applied Statistics with S-PLUS, 4th edn. Springer, New York (2002)

[60] van der Waerden, B.: Mathematische Statistik. Springer, Berlin (1957)

[61] Wets, R.J.B.: Statistical estimation from an optimization viewpoint. Ann. Oper. Res. **85**, 79–101 (1999)

[62] Yaari, M.E.: The dual theory of choice under risk. Econometrica **55**(1), 95–115 (1987)

[63] Zabarankin, M., Pavlikov, K., Uryasev, S.: Capital asset pricing model (CAPM) with drawdown measure. Eur. J. Oper. Res. (2013, to appear)

Index

M. Zabarankin and S. Uryasev, *Statistical Decision Problems: Selected Concepts and Portfolio Safeguard Case Studies*, Springer Optimization and Its Applications 85, DOI 10.1007/978-1-4614-8471-4,

MIX
Papier aus verantwortungsvollen Quellen
Paper from responsible sources
FSC® C105338

If you have any concerns about our products,
you can contact us on
ProductSafety@springernature.com

In case Publisher is established outside the EU,
the EU authorized representative is:
Springer Nature Customer Service Center GmbH
Europaplatz 3, 69115 Heidelberg, Germany

Printed by Libri Plureos GmbH
in Hamburg, Germany